LIFE-CYCLE COST MODELS FOR GREEN BUILDINGS

LIFE-CYCLE COST MODELS FOR GREEN BUILDINGS

With Optimal Green Star Credits

I.M. CHETHANA S. ILLANKOON

Associate Lecturer, School of Architecture and Built Environment, University of Newcastle, Callaghan, NSW, Australia

VIVIAN W.Y. TAM

Professor, School of Built Environment, Western Sydney University, Penrith, NSW, Australia

KHOA N. LE

Associate Professor, School of Engineering, Western Sydney University, Penrith, NSW, Australia

ELSEVIER

Butterworth-Heinemann
An imprint of Elsevier

Butterworth-Heinemann is an imprint of Elsevier
The Boulevard, Langford Lane, Kidlington, Oxford OX5 1GB, United Kingdom
50 Hampshire Street, 5th Floor, Cambridge, MA 02139, United States

Library of Congress Cataloging-in-Publication Data
A catalog record for this book is available from the Library of Congress

British Library Cataloguing-in-Publication Data
A catalogue record for this book is available from the British Library

ISBN: 978-0-12-820062-9

Publisher: Matthew Deans
Editorial Project Manager: Emily Thomson
Production Project Manager: Kamesh Ramajogi
Cover Designer: Matthew Limbert

Typeset by SPi Global, India

Contents

List of figures

List of tables

List of equations

Abstract

The green building is a widely discussed topic worldwide as a solution to increasing adverse impacts on the environment. The paradigm shift from conventional to green buildings is expected to yield environmental, social, and economic benefits. However, green building implementation is adversely affected by initial cost premiums although there are significant savings throughout the life-cycle of green buildings in terms of water, energy, and so on. Therefore, there is a clear need to analyze the initial stages of green building development regarding life-cycle impacts, capturing massive savings in energy, water, and other resources.

Although it may be cheaper to select inappropriate technologies during the initial decision-making stages, more importantly, this may preclude life-cycle savings and the desired outcomes of green buildings. In order to aid the initial decision-makers with the selection of credit points considering the lowest life-cycle costs of green buildings, this book presents a life-cycle cost model that incorporates developer constraints while maximizing the number of credit points achieved when using the Green Star Australia environmental rating system.

The model is based on Green Star Design and As-Built version 1.1 rating tool. Initially, an extensive analysis is carried out for all the key criteria and credits of Green Star Design and As-Built version 1.1 rating tool. Based on the identification of different types of credits, certain credits were eliminated. Afterwards, interdependencies among various credits were established. For all the selected credits, life-cycle cost is calculated considering six main central business districts (CBDs) of Australia. The life-cycle cost calculation followed "Building and construction assets—Service life planning—Part 5: Life-cycle costing standard" published by the International Organization for Standards (ISO) as a guideline. The net present value (NPV) technique is used to calculate life-cycle costs. Further, a sensitivity analysis is also carried out for selected credits to identify the changes to life-cycle cost to the changes in discount rate. Once all the life-cycle cost data is calculated, the proposed model was developed.

The proposed model is developed considering a set of rules for exclusions, selections, and inter-dependencies. It initially collects user information and user constraints. Based on the user information, the model provides customized solutions to the users. The user can define the discount rate and even select the regional areas, and based on that information, the life-cycle cost is calculated by the proposed model. The user constraints select or eliminate credits, consider inter-dependencies, and calculate the optimum solutions for a specific green certification level. This model can provide optimum solutions for four-star or five-star certification levels considering Green Star rating.

Finally, the proposed life-cycle cost model is validated in terms of cost and optimum credit selections. Cost is validated using costs comparisons with cost databases, industry reports, and actual green-certified buildings and interviews. To validate the credit selections, four case study buildings with Green Star certifications are considered. Based on the validation results, the cost calculations are within the range accepted by various sources. Further, the optimum credits proposed by the life-cycle cost model coincide with the credits obtained by the certified green buildings except for minor changes. Most of the credits that are proposed by the model yet not implemented by the case study buildings happen to have higher initial costs and lower life-cycle costs. This further strengthens the importance of using life-cycle costs during the initial decision-making stages for green building implementation. Further, credits with lower life-cycle costs are mostly eliminated owing to higher initial costs, which can be addressed by using the proposed life-cycle cost model.

The model identified green building credits with cost savings, such as the use of photovoltaic panels, which are ignored during the initial stages owing to high initial costs. Further, this model proposed passive methods such as natural ventilation in buildings, using daylight and rainwater tanks to be considered for green building implementations. Out of all the key criteria in Green Star Design and As-Built v1.1, credits representing management criterion are widely achieved in green building implementation. This perfectly coincides with the proposed life-cycle cost model.

Acknowledgment

It is with great pleasure that we offer our acknowledgment and gratitude to all the individuals who helped us in various ways to make this book a success. Many names will always remain in our minds and richly deserve our heartfelt gratitude.

We wish to acknowledge the financial support provided by the Australian Research Council (ARC) and the Australian Government (No: DP150101015; DP190100559).

Further, we also extend our sincere gratitude to Mr. Sajith Vimukthi Weerakoon and Mr. Jude Fernando for all their support. We are also grateful to all our friends at Western Sydney University, especially all the members of our research team.

List of abbreviations

Symbols

C_{sum}	sum of all the credit points of the given criteria
CO_2	carbon dioxide
CO_2-eq	carbon dioxide equivalent
E_{sum}	total life-cycle cost for energy credits
Em_{sum}	total life-cycle cost for emissions credits
i	discount rate
IEQ_{sum}	total life-cycle cost for indoor environment quality credits
L_{sum}	total life-cycle cost for land use and ecology credits
LCC_{total}	location-adjusted total life-cycle cost
M_{sum}	total life-cycle cost for managements credits
Mat_{sum}	total life-cycle cost for material credits
N	total number of periods
R_m	annual maintenance cost
R_t	net cash flow
t	time of cash flow
T_{sum}	total life-cycle cost for transport credits
W_{sum}	total life-cycle cost for water credits
x	initial cost per square meter for each credit point of a given criterion
y	present values of maintenance replacement and other costs
z	present value of demolition costs

Abbreviations

ABGR	Australian Building Greenhouse Rating
ASHRAE	American Society of Heating, Refrigerating and Air Conditioning Engineers
ARC	Australian Research Council
AUD	Australian Dollars
BASIX	building sustainability index
BCA	building construction authority
BD+C	building design and construction
BEAM	building environmental assessment method
BEE	building environment efficiency
BRE	building research establishment
BREEAM	Building Research Establishment Environmental Assessment Method
BSRIA	Building Services Research Information Association
CASBEE	Comprehensive Assessment System for Building Environmental Efficiency
CBD	central business district
CH2	Council House 2
CIB	International Council for Research and Innovation in Building Construction
DHW	domestic hot water
E	energy criterion
Em	emission criterion

EMP	environmental management plan
EPD	environmental product declarations
EVM	earned value management
GBCA	Green Building Council of Australia
GBI	Green Building Index
GBTool	green building tool
GHG	green house gases
GFA	gross floor area
GST	Goods and Services Tax
HK-BEAM	Hong-Kong Building Environmental Assessment Method
HVAC	heating, ventilating and air conditioning
I	innovation criterion
IAQ	indoor air quality
ID+C	interior design and construction
IEQ	indoor environment quality
IGBC	Indian Green Building Council
ISO	International Organization for Standards
IUCN	International Union for Conservation of Nature and Natural Resources
L	land use and ecology criterion
L_{BE}	built environment load
LCGCA	life cycle green cost assessment
LEED	Leadership in Environmental and Energy Design
M	management criterion
Mat	material criterion
MWR	multi-water reuse
NABERS	National Australian Built Environment Rating System
NatHERS	Nationwide House Energy Rating Scheme
NC	new construction
ND	neighborhood development
NPV	net present value
NR	non-residential
NSW	New South Wales
ODP	ozone depletion potential
O+M	operations and maintenance
OWM	Operational Waste Management
PAM	Malaysian Institute for Architects
PCM	phase change material
PV	photovoltaic
PVA	present value of annuity
PVC	poly vinyl chloride
Q	built environment quality
RI	regional index
SCM	supplementary cementitious material
T	transport criterion
UK	United Kingdom
USA	United States of America
USGBC	United States Green Building Council
VOC	volatile organic compound
W	water criterion
WELS	water efficiency labelling system
WGBC	World Green Building Council

CHAPTER

1

Introduction

1.1 Background

Intensive increase in human demands has been the subject of many adverse environmental conditions and this phenomenon has been widely reported as a cause of various catastrophes. By now, it is not possible to continue the pace of human development without facing environmental consequences (Warren, 2010). Therefore, to strike a balance between environmental impacts and the development, man is now looking for many solutions for the actions of the past. As a result, "sustainable development," and "green buildings" have become buzzwords in this era. This scope of interest renders green buildings a worthwhile research interest.

As a result of many environmental and social impacts, sustainable development has come into light. There are many definitions and illustrations put forward for "sustainable development" over the past couple of decades (Dincer & Rosen, 1999; Lele, 1991; McDonald, 1996; Mebratu, 1998; Mitcham, 1995; Omer, 2008; Parkin, 2000). According to Heinburg (2010), this term gained much wider usage after the publication of the Brundtland (1987) Report from the United Nations' World Commission on Environment and Development. According to the Brundtland Report in 1987, sustainable development was identified as the usage of resources in such a way without depriving future generations of benefitting from those resources (Brundtland, 1987).

Many commentators expand the vision of Brundtland on sustainable development into three subordinate concepts: social, economic, and environmental sustainability (Carew & Mitchell, 2008). Usually, environmental sustainability refers to natural resources, social sustainability is the need to sustain the resources, and economic sustainability is the mechanism used in meeting those needs (Mitchell, 2000). Further, in the Johannesburg Declaration (United Nations, 2002), socio-economic and environmental targets were set, and that created a collective responsibility to advance and strengthen these three pillars (Robert, Parris, & Leiserowitz, 2005).

https://doi.org/10.1016/B978-0-12-820062-9.00001-3

However, environmental sustainability is more challenged by the construction practices (Yılmaz & Bakış, 2015). Many studies are available that confirm that construction activities have significant negative impacts toward the environment (Abidin, 2010; Tan, Shen, & Yao, 2011). The building sector is one of the significant sectors that emits greenhouse gases (GHG) in Australia (Reidy, Lederwasch, & Ison, 2011) and globally, which leads to detrimental environmental impacts. This fact is statistically proven in the report by the United States Green Building Council Research Committee (2008) which indicated that buildings in the United States of America (USA) were responsible for about 38% of CO_2 emissions, about 71% of electricity consumption, about 39% of energy use, about 12% of water consumption, and about 40% of non-industrial waste. Further, minimizing the carbon footprint of the building can lead to a significant positive impact on the global environment (Green Building Council of Australia, 2013a). These facts and figures significantly illustrate the contribution of the building sector toward this grave situation. Therefore, the construction industry is always challenged to cater its demand in a socially and ecologically responsible way (DuPlessis, 2007). As a result, the green building concept came in to light.

There are many definitions and illustration of green buildings available in the literature. Richardson and Lynes (2007) identified "green buildings" as more energy- and resource-efficient compared to standard buildings. Green Building Council of Australia (2018) identified a green building as one that incorporates design, construction, and operational practices that significantly reduce or eliminate its negative impact on the environment and its occupants. With all the highlighted environmental impacts and the contribution of the construction industry to worsening the situation, the green building concept has gained its momentum (Hoffman & Henn, 2008). Further, green buildings must not be considered as a choice or a luxury but an essential for an environmentally concerned society (Ashuri & Durmus-Pedini, 2010).

In 2011, the General Services Administration of the USA selected 22 green buildings to form the national portfolio and derived remarkable results on the performance of green buildings (General Service Administration USA, 2011). According to the study, green buildings require approximately 25% less energy and 19% lower aggregate operational costs and enjoy 27% higher occupant satisfaction and 36% fewer CO_2 emissions compared to the national standards. Therefore, it is evident that many environmental and social benefits can be derived from green buildings.

Green building implementation is one of the most widely discussed topics of this decade. However, green building implementation is strongly associated with higher costs. There are mixed reviews in the literature regarding the initial cost (Richardson & Lynes, 2007). According to Tatari and Kucukvar (2011), LEED registered buildings have to pay a cost premium of 0.66% to obtain a certified status, 2.11% for silver status, 4.41% for gold status, and 6.5% for platinum status. Further, Dwaikat and Ali (2016) illustrated that the maximum reported cost premium of a green building reaches up to 21%. However, these studies focused on the initial cost of the green building.

In contrast, Bordass (2000) argued that the capital cost in initial stages can be dwarfed by the lower operating and maintenances cost. According to the World Green Building Council (2018), a green building is expected to incur an increase of 0.4–12.5% in design and construction costs while experiencing an energy saving of up to 25–30%.

Following a thorough investigation of 428 Green Star certified projects, it is reported that on average, Green Star certified buildings use 66% less electricity and 51% less potable water than conventional Australian buildings (Green Building Council of Australia, 2013b). Similarly, according to McGraw Hill Construction (2013), in new green buildings, operation costs decrease by more than 8% over a period of 1 year, while green retrofits exhibit a decrease of 9%. For example, certain green buildings were reported to consume 26% less energy and demonstrated 13% lower maintenance costs compared to average commercial buildings (Fowler & Rauch, 2006). These cost savings throughout the life-cycle are not captured in the initial cost calculations. The report of California's sustainable building task force clearly illustrated that higher initial costs for green buildings are a societal perception due to the lack of life-cycle costing in practice (Kats, Alevantis, Berman, Mills, & Perlman, 2003). Similarly, Tam, Hao, and Zeng (2012) illustrated that lack of knowledge of life-cycle cost as a factor which adversely affects the implementation of green buildings as usually, the initial cost is higher for green buildings, although maintenance cost should significantly incur savings. Therefore, it is necessary to develop a mechanism to capture these savings during the initial building stages. Similarly, Zhang, Wu, and Liu (2018), illustrated that the life-cycle perspective is overlooked in green buildings and therefore, there is an urgent need to provide comprehensive and urgent evidence against life-cycle costs.

Life-cycle costing is a method used to predict and assess the cost performance of assets (International Organisation for Standardization [ISO], 2017). In the construction industry, various research studies exist on identifying the life-cycle costs of specific material applications and systems separately (Illankoon, Tam, Le, & Wang, 2018; Tam, Senaratne, et al., 2017; Tam, Le, Zeng, Wang, & Illankoon, 2017). Further, there are researches considering optimum solutions for different green building solutions focusing the life-cycle perspective such as heating, cooling, and ventilation systems, (Johansson, 2009), timber applications (Tam, Senaratne, et al., 2017), vertical greening systems (Perini & Rosasco, 2016), optimal thermal comfort designs (Kim, Hong, Jeong, Koo, & Jeong, 2016), solar panels (Tam, Le, et al., 2017), green roofs (Sangkakool & Techato, 2016), roof top gardens (Wong, Tay, Wong, Ong, & Sia, 2003), single family detached houses (Hasan, Vuolle, & Sirén, 2008), wall material (Emmanuel, 2004), transparent insulation facades (Wong, Perera, & Eames, 2010), flooring (Allacker, 2012; Minne & Crittenden, 2015), and so on. All of these life-cycle cost studies analyzed specific requirements in green buildings, completely ignoring the other considerations.

Similarly, there are optimization models developed to identify the optimum solutions considering the lowest life-cycle cost and one specific criterion of green buildings, such as energy efficiency (Mithraratne & Vale, 2004; Verbeeck & Hens, 2007), water efficiency (Chai, Hu, Peng, & Wang, 2010), air conditioning usage (Bichiou & Krarti, 2011; Hasan et al., 2008), and so on. Once again, these life-cycle cost optimization models also considered only one specific criterion whereas a green building is a combination of various criteria with many green initiatives. Park, Choi, Kim, Jeong, and Kong (2017) developed an optimization model for LEED-certified buildings considering initial costs. In this model, the life-cycle cost savings are not captured.

Green building implementation has become a requirement in the status quo and therefore, the negative influence of initial cost premiums needs to be eradicated from the green building concept. For that purpose, in the initial decision-making stages there should be a model to look into the life-cycle costs of green buildings considering all the life-cycle savings. According to Zuo et al. (2017), the uptake of life-cycle cost in construction is rather slow in the Australian context. Further, according to literature, there is a lack of reliable life-cycle cost data and life-cycle cost calculation models to clients and designers in pondering different green building options to achieve the overall goal of green buildings. Therefore, this book aims to propose a life-cycle cost model to select the optimum credits in commercial office buildings in Australia.

1.2 Aim and objectives

The aim of this book is to develop a life-cycle cost model for optimum selection of Green Star credits for commercial office buildings in Australia. To fulfill this aim, the following objectives were established.

1. Critical review of the literature related to:
   a. Sustainable development and construction
   b. Green buildings and green building rating tools
   c. Green Star rating tools in Australia
   d. Initial cost premium of green buildings
   e. Life-cycle cost for green buildings
   f. Life-cycle cost models developed for green buildings

2. Examining Green Star Design and As-Built v1.1. This includes identifying credits of Green Star rating system. There are two parts to this:
   a. Classification of credits
   b. Identifying the interdependencies among the credits
3. Establishing life-cycle costs for all the selected credits in Green Star Design and As-Built v1.1. This objective is achieved after achieving the following:
   a. Identifying the different cost components associated with each of the credits
   b. Collecting the relevant cost data and relevant maintenance data
   c. Developing and identifying sensible assumptions for life-cycle cost calculations, if required
4. Developing a life-cycle cost model for optimum selection of Green Star credits for commercial office buildings in Australia. To develop the proposed life-cycle cost model, the following are investigated:
   a. Defining the user inputs
   b. Defining the dependencies, exclusions, and selection rules
   c. Developing the user interface
5. Validation of proposed life-cycle cost model for optimum selection of Green Star credits for commercial office buildings in Australia. The validation is carried out for the following sections
   a. Validating the life-cycle costs calculations
   b. Validating the proposed model using case studies

1.3 Significance of the research study

As illustrated in the background, there are many limitations associated with the available life-cycle cost models. Most of the life-cycle cost models focus on a separate criterion for green buildings and therefore, there is a clear lack of a life-cycle cost model to address the green buildings holistically. However, green buildings should be environmentally, socially, and economically sustainable. Considering these triple bottom line constructs, there are many performance criteria embedded into green buildings, such as water efficiency, energy efficiency, sustainable sites IEQ, materials, and so on. Further, when evaluating the performance of green buildings using green building rating tools, these criteria are evaluated. Therefore, when developing a life-cycle cost model for green buildings, all these criteria must be embedded into the proposed model, which is clearly lacking in the current literature. The significance of this proposed life-cycle cost model is that it considers all the key criteria required for green buildings when providing optimum solutions considering the life-cycle cost. Therefore, clients, designers and policy makers can utilize this proposed life-cycle cost model to derive optimum solutions based on given user information and constraints.

1.4 Research methodologies

The research process for this research study consists of three phases and further broken down into logical steps as given in Fig. 1.1. Phase I of the book included the literature review and established the aim of the book. Phase two included collection of cost data, and selecting Green Star credits (refer Fig. 1.1). Calculation of life-cycle cost for the identified credits marked the end of phase II. Finally, phase III focused on the proposed life-cycle cost model development.

To achieve the aim and objectives, the following research methodologies will be adopted.

- Literature review

  An extensive literature review is carried out to identify the concepts of sustainable development, sustainable construction, green buildings, green building rating tools, Green Star rating tool Australia, costs of green buildings, life-cycle of green buildings, and life-cycle cost model developed for green buildings.
- Life-cycle cost calculation

  Life-cycle cost calculation is carried out with the use of estimating techniques, relevant assumptions, and cost data. The net present value (NPV) technique is used to arrive at the present value of costs incurred within the operational stage of the life-cycle. There are certain underlying sensible assumptions made for the life-cycle cost calculations. These are illustrated in the latter sections of this book and given in the proposed model where necessary.

After calculating the life-cycle cost for each credit, there is a regional index included, changing the cost if the building is located in a regional area. All the life-cycle costs calculations are carried out for main six central business districts (CBDs) in Australia.

- Life-cycle cost model development

  The model is developed based on Java software. Based on the various interdependencies among credits, various rules are developed. The proposed model selects the optimum solutions with the lowest life-cycle and maximum total credits points.
- Proposed model validations

  The proposed model validation is twofold. Initially, the life-cycle cost data is validated. There are interviews carried out to validate the life-cycle costs. Aside from this, these costs are compared among different available composite cost data for validation.

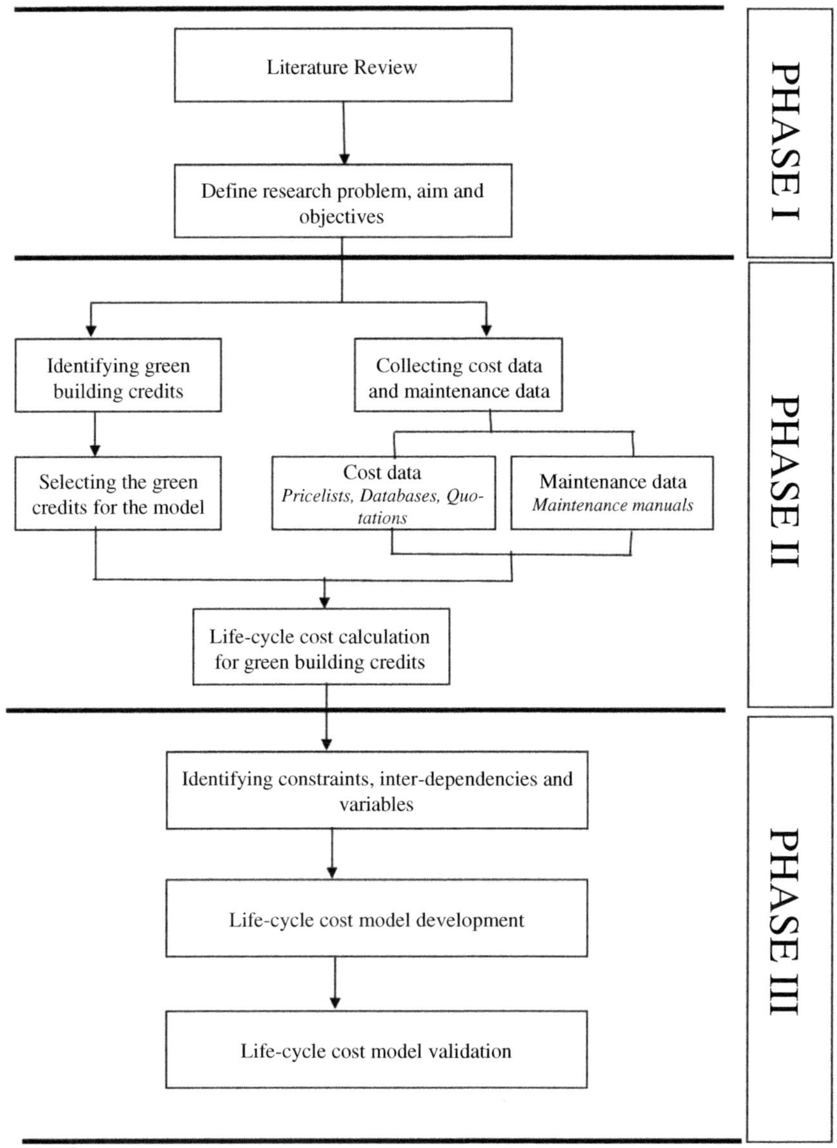

FIG. 1.1 Research process.

The proposed life-cycle cost model is also validated considering four case study office buildings with Green Star certification in Australia. The four case study buildings are located in four main CBDs in Australia. The building details and the selections of the proposed model are compared for validation.

1.5 Scope of the research

This book proposed a life-cycle cost model for green commercial office buildings in Australia. The proposed life-cycle cost model followed Green Star Design and As-Built v1.1 as the guideline for green building assessment. Further, the proposed model considered a 60 year life-cycle for green building for life-cycle cost calculations.

1.6 Structure of the book

Chapter one of the book illustrates the background to the research, including the aim, objectives, and a brief illustration of the methodologies used to carry out the research.

The **second chapter** focuses on the basic concepts of sustainable development and sustainable constructions. Further, it provides a detailed illustration of green buildings and the triple bottom line concept. This chapter further identifies the green building rating tools used worldwide, then provides basic details of eight widely used green building rating tools representing the main five geographical regions. There is a separate section of this chapter illustrating the Green Star green building rating tool in Australia.

The third chapter of the book focuses on the cost of green buildings and compares the initial cost of green buildings with conventional buildings. This chapter illustrates the available life-cycle cost model developed for green buildings.

The fourth chapter focuses on the Green Star Design and As-Built version 1.1 credits. There is a detailed analysis on every credit included in the Green Star rating tool. Afterwards, this chapter illustrates the interdependencies among various credits included in this book.

The fifth chapter of this book provides all the detailed information on life-cycle cost calculations for each credit. Initially, this chapter provides in-depth information on the life-cycle costing technique used in the book and defines the various cost components included in the life-cycle cost calculation. Life-cycle cost calculations for each key criterion are given in separate sections. All the sensible assumptions are illustrated in this chapter. Further, various cost components included in each credit are also provided. Finally, this chapter provides a sensitivity analysis of the changes in life-cycle costs to the change in discount rate.

The sixth chapter focuses on the model development. In this chapter, there is a detailed illustration of different exclusion rules, dependencies, and selection rules used to develop the model. Further, it provides information on user inputs and the algorithm used to develop the proposed life-cycle cost model.

The seventh chapter of this book illustrates the validation. The validation of the research is twofold. There are two separate sections in this chapter focusing on the life-cycle cost validation and the proposed model validation considering the four case study buildings.

The eighth chapter and final chapter of this book illustrates the main conclusions and recommendations from this book. Further, it includes the limitations and future research directions.

References

Abidin, N. Z. (2010). Investigating the awareness and application of sustainable construction concept by Malaysian developers. *Habitat International, 34*(4), 421–426. https://doi.org/10.1016/j.habitatint.2009.11.011.

Allacker, K. (2012). Environmental and economic optimisation of the floor on grade in residential buildings. *The International Journal of Life Cycle Assessment, 17*(6), 813–827. https://doi.org/10.1007/s11367-012-0402-2.

Ashuri, B., & Durmus-Pedini, A. (2010). An overview of the benefits and risk factors of going green in existing buildings. *International Journal of Facility Management, 1*(1).

Bichiou, Y., & Krarti, M. (2011). Optimization of envelope and HVAC systems selection for residential buildings. *Energy and Buildings, 43*(12), 3373–3382. https://doi.org/10.1016/j.enbuild.2011.08.031.

Bordass, B. (2000). Cost and value: Fact and fiction. *Building Research & Information, 28*(5–6), 338–352. https://doi.org/10.1080/096132100418492.

Brundtland, G. H. (1987). *Report of the world commission on environment and development: Our common future (A/42/427)*. Retrieved from Oslo: http://www.un-documents.net/our-common-future.pdf.

Carew, A., & Mitchell, C. (2008). Teaching sustainability as a contested concept: Capitalizing on variation in engineering educators' conceptions of environmental, social and economic sustainability. *Journal of Cleaner Production, 16*(1), 105–115. https://doi.org/10.1016/j.jclepro.2006.11.004.

Chai, H. X., Hu, X. B., Peng, S. J., & Wang, T. Y. (2010). Incremental-cost economy model in full life cycle of water-saving projects for green building. *Huanan Ligong Daxue Xuebao/Journal of South China University of Technology (Natural Science), 38*(11), 59–63. https://doi.org/10.3969/j.issn.1000-565X.2010.11.011.

Dincer, I., & Rosen, M. A. (1999). Energy, environment and sustainable development. *Applied Energy, 64*(1), 427–440. https://doi.org/10.1016/S0306-2619(99)00111-7.

DuPlessis, C. (2007). A strategic framework for sustainable construction in developing countries. *Construction Management and Economics, 25*(1), 67–76. https://doi.org/10.1080/01446190600601313.

Dwaikat, L. N., & Ali, K. N. (2016). Green buildings cost premium: A review of empirical evidence. *Energy and Buildings, 110*(Suppl. C), 396–403. https://doi.org/10.1016/j.enbuild.2015.11.021.

Emmanuel, R. (2004). Estimating the environmental suitability of wall materials: Preliminary results from Sri Lanka. *Building and Environment, 39*(10), 1253–1261. https://doi.org/10.1016/j.buildenv.2004.02.012.

Fowler, K. M., & Rauch, E. M. (2006). *Sustainable building rating systems summary (PNNL-15858)*. Retrieved from http://www.pnl.gov/main/publications/external/technical_reports/PNNL-15858.pdf.

General Service Administration USA. (2011). *Assessing green building performance; A post occupancy evaluation of 12 GSA buildings*. Retrieved from http://www.gsa.gov/graphics/pbs/GSA_AssessGreen_white_paper.pdf.

Green Building Council of Australia. (2013a). *Evolution: A year in sustainable building*. Retrieved from https://www.gbca.org.au/resources/gbca-publications/green-building-evolution-2013/.

Green Building Council of Australia. (2013b). *The value of Green Star—A decade of environmental benefits*. Retrieved from https://www.gbca.org.au/uploads/194/34754/The_Value_of_Green_Star_A_Decade_of_Environmental_Benefits.pdf.

Green Building Council of Australia. (2018). *About GBCA*. Retrieved from http://www.gbca.org.au/about/.

Hasan, A., Vuolle, M., & Sirén, K. (2008). Minimisation of life cycle cost of a detached house using combined simulation and optimisation. *Building and Environment*, *43*(12), 2022–2034. https://doi.org/10.1016/j.buildenv.2007.12.003.

Heinburg, R. (2010). *What is sustainability?* Santa Rosa, CA: Post Carbon Institute.

Hoffman, A. J., & Henn, R. (2008). Overcoming the social and psychological barriers to green building. *Organization & Environment*, *21*(4), 390–419. https://doi.org/10.1177/1086026608326129.

Illankoon, I. M. C. S., Tam, V. W. Y., Le, K. N., & Wang, J. Y. (2018). Life cycle costing for obtaining concrete credits in green star rating system in Australia. *Journal of Cleaner Production*, *172*, 4212–4219. https://doi.org/10.1016/j.jclepro.2017.11.202.

International Organisation for Standardization [ISO]. (2017). *Buildings and constructed assets—Service life planning Part 5: Life cycle costing (Vol. ISO 15686-5)*. Switzerland: International Standard Organisation [ISO].

Johansson, D. (2009). The life cycle costs of indoor climate systems in dwellings and offices taking into account system choice, airflow rate, health and productivity. *Building and Environment*, *44*(2), 368–376. https://doi.org/10.1016/j.buildenv.2008.03.011.

Kats, G., Alevantis, L., Berman, A., Mills, E., & Perlman, J. (2003). *The costs and financial benefits of green buildings. A report to California's sustainable building task force. USA.*

Kim, J., Hong, T., Jeong, J., Koo, C., & Jeong, K. (2016). An optimization model for selecting the optimal green systems by considering the thermal comfort and energy consumption. *Applied Energy*, *169*, 682–695. https://doi.org/10.1016/j.apenergy.2016.02.032.

Lele, S. M. (1991). Sustainable development: A critical review. *World Development*, *19*(6), 607–621. https://doi.org/10.1016/0305-750X(91)90197-P.

McDonald, G. T. (1996). Planning as sustainable development. *Journal of Planning Education and Research*, *15*(3), 225–236. https://doi.org/10.1177/0739456X9601500306.

McGraw Hill Construction. (2013). *World green building trends*. Retrieved from http://www.worldgbc.org/files/8613/6295/6420/World_Green_Building_Trends_SmartMarket_Report_2013.pdf.

Mebratu, D. (1998). Sustainability and sustainable development: Historical and conceptual review. *Environmental Impact Assessment Review*, *18*(6), 493–520. https://doi.org/10.1016/S0195-9255(98)00019-5.

Minne, E., & Crittenden, J. C. (2015). Impact of maintenance on life cycle impact and cost assessment for residential flooring options. *The International Journal of Life Cycle Assessment*, *20*(1), 36–45.

Mitcham, C. (1995). The concept of sustainable development: Its origins and ambivalence. *Technology in Society*, *17*(3), 311–326. https://doi.org/10.1016/0160-791X(95)00008-F.

Mitchell, C. (2000). Integrating sustainability in chemical engineering practice and education: Concentricity and its consequences. *Process Safety and Environmental Protection*, *78*(4), 237–242. https://doi.org/10.1205/095758200530754.

Mithraratne, N., & Vale, B. (2004). Life cycle analysis model for New Zealand houses. *Building and Environment*, *39*(4), 483–492. https://doi.org/10.1016/j.buildenv.2003.09.008.

Omer, A. M. (2008). Energy, environment and sustainable development. *Renewable and Sustainable Energy Reviews*, *12*(9), 2265–2300. https://doi.org/10.1016/j.rser.2007.05.001.

Park, J.-Y., Choi, S.-G., Kim, D.-K., Jeong, M.-C., & Kong, J.-S. (2017). Credit optimization algorithm for calculating LEED costs. *Sustainability*, *9*(9), 1607.

Parkin, S. (2000). Sustainable development: The concept and the practical challenge. In *Paper presented at the Proceedings of the ICE-Civil Engineering*.

Perini, K., & Rosasco, P. (2016). Is greening the building envelope economically sustainable? An analysis to evaluate the advantages of economy of scope of vertical greening systems and green roofs. *Urban Forestry & Urban Greening*, *20*, 328–337. https://doi.org/10.1016/j.ufug.2016.08.002.

Reidy, C., Lederwasch, A., & Ison, N. (2011). *Defining zero emission buildings review and recommendations: Final report*. Retrieved from http://www.asbec.asn.au/research/.

Richardson, G. R., & Lynes, J. K. (2007). Institutional motivations and barriers to the construction of green buildings on campus: A case study of the University of Waterloo, Ontario. *International Journal of Sustainability in Higher Education*, *8*(3), 339–354. https://doi.org/10.1108/14725960410808096.

Robert, K. W., Parris, T. M., & Leiserowitz, A. A. (2005). What is sustainable development? Goals, indicators, values, and practice. *Environment: Science and Policy for Sustainable Development*, *47*(3), 8–21. https://doi.org/10.1080/00139157.2005.10524444.

Sangkakool, T., & Techato, K. (2016). Life cycle cost of air plant green roofs in hot and humid climate. *International Journal of Applied Business and Economic Research*, *14*(10), 7167–7182.

Tam, V. W. Y., Hao, J. L., & Zeng, S. (2012). What affects implementation of green buildings? An empirical study in Hong Kong. *International Journal of Strategic Property Management*, *16*(2), 115–125. https://doi.org/10.3846/1648715X.2011.645559.

Tam, V. W. Y., Le, K. N., Zeng, S. X., Wang, X., & Illankoon, I. M. C. S. (2017). Regenerative practice of using photovoltaic solar systems for residential dwellings: An empirical study in Australia. *Renewable and Sustainable Energy Reviews*, *75*, 1–10. https://doi.org/10.1016/j.rser.2016.10.040.

Tam, V. W. Y., Senaratne, S., Le, K. N., Shen, L.-Y., Perica, J., & Illankoon, I. M. C. S. (2017). Life-cycle cost analysis of green-building implementation using timber applications. *Journal of Cleaner Production*, *147*, 458–469. https://doi.org/10.1016/j.jclepro.2017.01.128.

Tan, Y., Shen, L., & Yao, H. (2011). Sustainable construction practice and contractors' competitiveness: A preliminary study. *Habitat International*, *35*(2), 225–230. https://doi.org/10.1016/j.habitatint.2010.09.008.

Tatari, O., & Kucukvar, M. (2011). Cost premium prediction of certified green buildings: A neural network approach. *Building and Environment*, *46*(5), 1081–1086. https://doi.org/10.1016/j.buildenv.2010.11.009.

United Nations. (2002). *Johannesburg declaration on sustainable development (A/CONF.199/20)*. Retrieved from Johannesburg: http://www.joburg.org.za/pdfs/johannesburgdeclaration.pdf.

United States Green Building Council Research Committee. (2008). *A national green building research agenda*. Retrieved from http://www.usgbc.org/resources/national-green-building-research-agenda.

Verbeeck, G., & Hens, H. (2007). Life cycle optimization of extremely low energy dwellings. *Journal of Building Physics*, *31*(2), 143–177. https://doi.org/10.1177/1744259107079880.

Warren, C. (2010). Measures of environmentally sustainable development and their effect on property asset value: An Australian perspective. *Property Management*, *28*(2), 68–79.

Wong, I. L., Perera, S., & Eames, P. C. (2010). Goal directed life cycle costing as a method to evaluate the economic feasibility of office buildings with conventional and TI-façades. *Construction Management and Economics*, *28*(7), 715–735. https://doi.org/10.1080/01446191003753867.

Wong, N. H., Tay, S. F., Wong, R., Ong, C. L., & Sia, A. (2003). Life cycle cost analysis of rooftop gardens in Singapore. *Building and Environment*, *38*(3), 499–509. https://doi.org/10.1016/S0360-1323(02)00131-2.

World Green Building Council. (2018). Retrieved from http://www.worldgbc.org/index.php?cID=1.

Yılmaz, M., & Bakış, A. (2015). Sustainability in construction sector. *Procedia—Social and Behavioral Sciences*, *195*, 2253–2262. https://doi.org/10.1016/j.sbspro.2015.06.312.

Zhang, L., Wu, J., & Liu, H. (2018). Turning green into gold: A review on the economics of green buildings. *Journal of Cleaner Production*, *172*, 2234–2245. https://doi.org/10.1016/j.jclepro.2017.11.188.

Zuo, J., Pullen, S., Rameezdeen, R., Bennetts, H., Wang, Y., Mao, G., … Duan, H. (2017). Green building evaluation from a life-cycle perspective in Australia: A critical review. *Renewable and Sustainable Energy Reviews*, *70*, 358–368. https://doi.org/10.1016/j.rser.2016.11.251.

CHAPTER

2

Sustainable development and green buildings

2.1 Introduction

This chapter consists of four sections. The first three sections discuss the basic concepts of sustainable development, sustainable construction and green buildings. The emergence of sustainable development as a concept is first discussed in brief, after which the term is defined. Next, sustainable construction is explained, and its key aspects are discussed in depth. The next section examines green buildings, and its key concepts in detail, focusing on various green building rating tools used worldwide. The Green Star rating tool in Australia is also explained in detail. Finally, this chapter analyses how each of the rating tools achieves the triple bottom-line of sustainability using various key criteria.

2.2 Sustainable development

Over the last few decades, *"Sustainability"* has emerged as a vibrant field of research. This concept evolved as a response to the limited availability of resources. This term is relative, and dates back to several years back into early human civilization. However, in recent times, the word *"sustainability"* has been used in many disciplines, and has also been misinterpreted in many instances (Kuhlman & Farrington, 2010).

Sutton (2004), discussed the origin of the word *"sustainability"* and argued that it evolved from the word "sustain," which was derived from the Latin word *"sustenare"* which means, "to hold up." However, the initial evidence of the term "sustainability" indicates that it was coined in the discipline of forestry (Kuhlman & Farrington, 2010; Sutton, 2004) in 1713, in a book called "Sylvicultura Occonomica," written by a German forester and scientist, Hans Carl von Carlowitz. At this point in time, the concept was discussed at a micro level. With massive environmental issues giving

https://doi.org/10.1016/B978-0-12-820062-9.00002-5

rise to the need to preserve the environment in contemporary times, sustainability has been developed into a macro concept (Sutton, 2004). After the concept was extended to the macro context, the term "sustainability" has often been discussed as "sustainable development," whereas in 1980, the World Conservation Strategy as published by International Union for Conservation of Nature and Natural Resources (IUCN), aimed at achieving sustainable development through the conservation of living resources.

However, the concept of sustainable development gained momentum with the Brundtland Report (Brundtland, 1987). In this report, keen consideration was given to several grave problems such as deforestation, ozone depletion, the greenhouse effect, and desertification. In Chapter 2 of the report, the term "sustainable development" was defined as:

> development that meets the needs of the present without compromising the ability of future generations to meet their own needs.

This definition of sustainable development was adopted by many research studies afterwards (Curran, 2009; Giddings, Hopwood, & O'brien, 2002; Kuhlman & Farrington, 2010). According to Grierson (2009), the Brundtland definition implies that a balance can be achieved between human socioeconomic activities, as well as natural resource availability and environmental preservation. The Brundtland definition gained several dimensions and expanded with time. Since its inception, as many as over 500 definitions of sustainable development were spawned by various governments, professional bodies, institutions, and organizations (Shah, 2008).

Dovers and Handmer (1992) illustrated sustainable development as the ability of a human system, natural or mixed, to resist or adapt to endogenous or exogenous change indefinitely, and, in addition, as a way of intentional change and improvement that keeps or increases this attribute of the system meeting the needs of the population. Compared to the Brundtland definition, Dovers and Handmer (1992) constantly discussed the changes and the ability to meet new demands, which has not been illustrated in the Brundtland definition.

According to Solow (1992), sustainable development is not only about preserving natural resources, but also about developing substitutions between natural and other sorts of capital. Similarly, Pirages (1994) also identified sustainable development as an evolving process and argued that sustainability should develop over time. Gladwin, Kennelly, and Krause (1995) also had similar views and identified sustainable development as a process of achieving human development in an inclusive, connected, equitable, prudent, and secure manner. In all these definitions, sustainable development was identified as a process which evolves to meet the changing needs of the population.

These definitions were strengthened by Vanegas, DuBose, and Pearce (1996), who explained that sustainable development is a dynamic concept rather than a static state, and that it required decision-makers to be flexible and willing to modify their approaches according to changes in the environment, human needs and desires, and technological advances. Similarly, sustainable development can also be discussed as development which improves the quality of human life while living within the carrying capacity of supporting eco-systems (Hill & Bowen, 1997). In these ideas on sustainable development, there was considerable focus on "development" as relevant to the time. Further, higher consideration was given towards the development of quality of human life, while striking a balance with the ecosystem.

Ross (2009) referred to sustainable development as things that can be done for a longer period of time without any unacceptable consequences. In a similar study, Ortiz, Castells, and Sonnemann (2009) identified sustainable development as a concept that enhances the quality of life, and therefore, allows people to live in a healthy environment and improves the social, economic, and environmental conditions for the present and future generations. While comparing these definitions with the Brundtland definition (Brundtland, 1987), the underlying basis can be identified as the same. In both cases, environmental concerns were given greater focus in terms of the future. Further, both definitions discuss the need to maintain the environment over time. However, in certain explanations on sustainable development, such as those by Dovers and Handmer (1992) and Ross (2009), there is greater focus on adaptation.

With the passing of time, this term will change its meaning to suit the situations to come. This fact was further strengthened by Wilbanks and Wilbanks (2010) who argued that challenges to sustainable development constantly evolve, as changes in conditions and driving forces emerge with little notice, and therefore, sustainable development is a process of adaptation. However, after thoroughly reviewing the literature on sustainable development and sustainability, Olawumi and Chan (2018) argued that although these two words are used interchangeably, sustainable development is a strategy, and sustainability refers to the process of achieving the strategy of sustainable development. However, it is believed that the most common and widely used definition of sustainability or sustainable development would be the definition provided by Brundtland (1987) (Olawumi & Chan, 2018).

In all these various definitions, two aspects have been considered, namely, preserving the environment and catering to the demand. There should be a balance between these two aspects. Therefore, in more common terms, the concept of sustainability has a threefold focus, namely, environmental, social and economic (Elkington, 1994), out of which environmental sustainability is considered the backbone of the other two.

In synthesizing all these definitions, Vos (2007) illustrated that nearly all shared core elements of sustainable development are related to economic, social and environmental considerations. According to Said and Berger (2013), sustainable development should be comprehensive and consider all the triple bottom-line aspects. These three domains are often identified as the three pillars of sustainable development, or the triple bottom-line (Carew & Mitchell, 2008; Kats, Alevantis, Berman, Mills, & Perlman, 2003), and therefore, it is illustrated as being the intersection of three overlapping circles.

According to Young (1997), economic, environment and social sustainability are like a three-legged stool. Therefore, if one leg is missing from the "sustainability stool," it will cause instability, because society, economy, and ecosystem are intricately linked together (Young, 1997). These three elements reinforce each other reciprocally, and, economic growth and social well-being are underpinned by environmental concerns, and vice versa (Svensson & Wagner, 2015). However, according to Welsford, See, and Erkki (2014), in order to sustain the environmental and social practices, the options must remain economically viable.

Social sustainability is concerned with the well-being of any person who is directly or indirectly affected by development efforts (Said & Berger, 2013). Thus, social well-being concerns human feelings, such as security, satisfaction, safety and comfort, and human contributions like skills, health, knowledge, and motivation (Parkin, 2000). According to Balkema, Preisig, Otterpohl, and Lambert (2002), the objective of social sustainability is to secure people's social-cultural and spiritual needs equitably, with stability in human morality, relationships, and institutions.

Environmental sustainability refers to the long-term viability of the natural environment, which is maintained to support long-term development by supplying resources and taking up emissions, and results in the protection and efficient utilization of environmental resources (Balkema et al., 2002). A much narrower explanation of this would be "not leading to the depletion of resources or the degradation of the environment." In other words, it would be overall viability and normal functioning of natural systems (Munasinghe, 2004). Sartori, Latrônico, and Campos (2014) identified environmental sustainability as the dematerialization of economic activity since a decrease in material processing can reduce the pressure on natural systems and expand the provision of environmental services to the economy.

Economic sustainability seeks to maximize the flow of income that could be generated while at least maintaining the stock of assets (or capital) which yields this income (Solow, 1992). Vanegas et al. (1996), explaining that economics, as it pertains to sustainability, does not simply refer to gross national product, exchange rates, inflation, and profit, but rather, has a broader meaning as being a social science that explains the production, distribution, and consumption of goods and services. This fact is further strengthened by Balkema et al. (2002), where economic sustainability should, in principle, include all resources, including those associated with social and environmental values. However, in practice, most analyses include only the financial costs and benefits. Further, growth, efficiency, and stability are identified as elements of economic sustainability (Munasinghe, 2004). While considering all these definitions, it is necessary to note that none of the triple bottom-line sustainability aspects can stand alone. The three are connected to each other, one way or the other, to achieve sustainable development. According to Olawumi and Chan (2018), these three triple bottom-line constructs should be harmonized to achieve holistic sustainable development.

However, the three same-sized overlapping circles for social, economic, and environmental sustainability is being criticized by Giddings et al. (2002) who argue that the model usually shows equal sized rings in a symmetrical interconnection, although there is no reason why this should be the case. Further, in their study, it was illustrated that in this model, it assumed the separation, and even autonomy, of the economy, society and environment from each other. Therefore, Giddings et al. (2002) proposed a different nested model for economic, social, and environmental sustainability.

According to Balkema et al. (2002), technical and cultural focus should be included in sustainable development. In the framework developed by United Nations Commission on Sustainable Development (2001), an institutional parameter was included as an indicator of sustainable development. Foxon et al. (2002) included a technical dimension to sustainable development. Further, it was explained that technical criteria would cover the reliability, durability, flexibility, and adaptability of a system. According to the Global Reporting Initiative (2011), the social dimension of the triple bottom-line was discussed in detail in terms of human rights, society, labor practices, and product responsibility. Although there has been a broad spectrum of literature on these three dimensions of sustainability, there have also been arguments to the effect that these three aspects alone do not solely cover the scope of sustainable development (Pawłowski, 2008). According to Pawłowski (2008) aside of the triple bottom-line dimension, moral, legal, technical, and political dimensions should be included in the sustainable development paradigm.

In summary, sustainable development is identified as environmental, social, and economic sustainability. These three pillars need to be addressed equally while addressing sustainable development. Further, with increments in human demands, the requirements for sustainable development have also expanded. Therefore, to cater to the expanded requirements, certain sub-categories need to be introduced to the sustainable development paradigm. Recent research studies such as those by

Balkema et al. (2002), United Nations Commission on Sustainable Development (2001), Foxon et al. (2002) and Pawłowski (2008), are set to identify these additions to the three pillars of sustainable development.

2.3 Sustainable development in construction

The construction industry plays a major role in any country's economy. Further, it has a significant impact on the environment, and social well-being (Baloi, 2003). Building construction around the world alone consumes about 40% of raw stone, gravel, and sand that is used each year, globally (World Watch Institute, 2015). About 25% of the total amount of virgin wood, about 40% of the world's energy and about 16% of the water used each year goes into building construction (World Watch Institute, 2015). Due to several adverse impacts as a consequence of all this use, it is necessary for the construction industry to move towards sustainable development. The sustainability performance of the construction industry signifies the overall degree to which the construction sector supports sustainable development in a particular economy (Ye, Zhu, Shan, & Li, 2015).

Sustainable construction is a separate discipline comprising various practical and theoretical frameworks (Hill & Bowen, 1997). In 1999, Agenda 21 for sustainable construction was published by the International Council for Research and Innovation in Building and Construction (1999). According to Agenda 21 for construction (International Council for Research and Innovation in Building and Construction, 1999), there are six principles to be adopted for sustainable construction, namely, maximization of resource reuse, minimization of resource consumption, use of renewable and recyclable resources, protection of the natural environment, creation of a healthy and non-toxic environment, and creation of quality in built environments. Further, Agenda 21 (International Council for Research and Innovation in Building and Construction, 1999), argues that initially, sustainability in construction is focused only on how to deal with limited resources, but with time, the scope has expanded and the emphasis has tended towards technical issues, such as material, building components, and construction technologies on energy-related design concepts. More recently, cultural issues and cultural heritage implications have gained more attention in sustainable construction (International Council for Research and Innovation in Building and Construction, 1999). This definition addressed all the triple bottom-line constructs of sustainable development (refer Section 2.2), except for the economic pillar. The first four principles address environmental sustainability and the latter two addresses economic sustainability.

However, according to Baloi (2003), sustainable construction addresses the triple bottom-line of sustainable development. Based on that study

(Baloi, 2003), the social dimension addresses issues pertaining to the enhancement of the quality of human life and the economic dimension addresses economics issues such as job creation, enhancement of competitiveness, lower operating or maintenance costs, high quality of working environment leading to greater productivity, and many others. Finally, the environmental dimension deals with the design, construction, operation, maintenance, and deconstruction approaches that minimize the adverse impacts of construction on the environment, in the form of emissions of air pollutants, waste discharge, use of water resources, land use, and others.

Further, Woolley, Kimmins, Harrison, and Harrison (2005) adopt the definition for sustainable construction provided by the Building Services Research and Information Association (BSRIA) as "the creation and responsible management of a healthy built environment based on resource-efficient and ecological principles." This definition is similar to the definition by the International Council for Research and Innovation in Building and Construction (1999), where the focus is on environmental and social parameters of sustainability.

According to Abidin (2010), sustainable construction is perceived as a way for the construction industry to contribute to the effort to reach sustainable development. Kibert (2012) illustrated that sustainable construction addresses the three pillars of sustainable development in the context of its community. Therefore, while considering all these definitions, most of them have put different requirements forward as a route to achieving sustainability in construction. These requirements fall into any of the three pillars of sustainability. However, certain definitions do not focus on economic sustainability, which is arguable.

A building has a long life-span. From the inception to its demolition, a building passes through different stages of a life-cycle. Considering this, Wyatt (1994) has deemed sustainable construction to include an appraisal, which includes managing the serviceability of a building during its lifetime, eventual deconstruction and recycling of resources, to reduce the waste stream that is usually associated with demolition. In 1994, in the conference proceedings of the International Council for Research and Innovation in Building and Construction (CIB), Kibert identified seven principles of sustainable construction (Kibert, 1994). These seven principles are: reduce resource consumption, reuse resources, use recyclable resources, protect nature, eliminate toxins, apply life costing, and focus on quality (Kibert, 1994). These principles can be applied across the entire life-cycle of construction, from planning to demolition. Similarly, DuPlessis (2007) outlined three aspects of sustainable construction. This first aspect is that sustainable construction requires a broad interpretation of construction as a process, involving many more players than just those that are traditionally identified as constituting the construction industry. Then, it emphasized both, environmental protection and value addition to the quality of life of individuals and communities. Finally, sustainable construction embraces not just technological responses, but also non-technical aspects related to social and economic sustainability. In this definition, the life-cycle of construction is considered, which is extremely significant.

Lavy and Fernández-Solis (2009) illustrated that the seven principles of Kibert (1994) consider carrying out the construction by reducing resource consumption, reusing resources, and using recycled resources, which tends to protect nature by eliminating toxins. Afterwards, Kibert (2012) explained that if these principles are adopted in developing a structure, the structure can be called a "*green building*."

2.4 Green buildings

Green buildings have become a widely discussed topic in recent times. There are many foundations, and governmental organizations that have been set up for the construction and development of green buildings. According to Cassidy, Wright, and Flynn (2003) the concept of "green buildings" emerged in the late 19th and early 20th Centuries. There are many studies that have focused on different aspects of green buildings, since then. However, by now, the concept has become a greater phenomenon.

Cassidy et al. (2003) identified green buildings as buildings which increase the efficiency of site, energy, water, and materials usage, and reduce building impacts on human health and the environment, through better siting, design, construction, operation, maintenance, and removal throughout the complete building life-cycle. Generally, buildings are designed to meet building code requirements, whereas green building design challenges designers to go beyond the codes to improve overall building performance, and minimize life-cycle environmental impact and cost (Gowri, 2004). Similarly, the ASHRAE (American Society of Heating, Refrigerating, and Air Conditioning Engineers) Guide (ASHRAE, 2006, p. 4) defined green design as:

> one that achieves high performance, over the full life-cycle, in the areas such as minimising natural resource consumption through more efficient utilisation of non-renewable natural resources, land, water, and construction materials, including utilisation of renewable energy resources to achieve net zero energy consumption, minimising emissions that negatively impact our indoor environment and the atmosphere of our planet, especially those related to indoor air quality (IAQ), greenhouse gases, global warming, particulates, or acid rain, minimising discharge of solid waste and liquid effluents, including demolition and occupant waste, sewer, and storm water, and the associated infrastructure required to accommodate removal, minimal negative impacts on site ecosystem, maximum quality of indoor environment, including air quality, thermal regime, illumination, acoustics/noise, and visual aspects to provide comfortable human physiological and psychological perceptions.

This definition is illustrative and provides ample information on the requirements for a green building. It discusses the minimisation of resource usage, advocates the use of renewable energy, providing better environments for occupants and environmental protection, and so on. Further, in this definition, the life-cycle of the green building is also considered. However, this definition does not include any inputs on economic considerations, which must be taken into account.

Similarly, the United States Green Building Council (2007) identified that a high performing green building is an efficient building with savings in energy costs ranging from 20% to 50%. It indicated that such buildings are created through integrated planning, site orientation, energy-saving technologies, on-site renewable energy-producing technologies, light-reflective materials, natural daylight and ventilation, and downsized heating, ventilation and air conditioning (HVAC), and other equipment. According to Hoffman and Henn (2008), "green building" is a term encompassing strategies, techniques, and construction products that are less resource-intensive or pollution-producing than products as a result of regular construction are. Similarly, according to Chan, Qian, and Lam (2009), green buildings bring together a vast array of practices and techniques to reduce the impacts of buildings on energy consumption, environment and human health.

The United States Environment Protection Agency (2014), identified green buildings as being a product of the practice of creating structures and using processes that are environmentally responsible and resource-efficient throughout the building's life-cycle, from siting to design, including construction, operation, maintenance, renovation, and deconstruction. Further, green buildings are designed to reduce the overall impact of the built environment on human health and the natural environment by efficiently using energy, water, and other resources, protecting occupant health, improving employee productivity, and reducing waste, pollution and environmental degradation (United States Environment Protection Agency, 2014). The World Wildlife Fund (2015) says that a green building is identified as being a physical structure that uses a design and planning process that is environmentally responsible and resource-efficient.

Considering all these explanations of green buildings, it is logical to identify a green building as an environmentally friendlier building with efficient use of energy, water, and other resources, providing a better living and working environment for its occupants. Further, a majority of the definitions focus on the health and well-being of the humans within the buildings (Cassidy et al., 2003; Chan et al., 2009; United States Green Building Council, 2007), while certain other definitions offer due consideration to the cost, as well (Gowri, 2004; United States Environment Protection Agency, 2014). While considering a green building, all the environmental, health and well-being, and economic aspects are considered throughout the life-cycle of the building, and not just for the construction phase alone.

In the literature, there are certain researchers who have used the terminology "green building" interchangeably with "sustainable building" (Nelms, Russell, & Lence, 2005). In the ASHRAE Guide (ASHRAE, 2006) these two are used interchangeably, and yet, it explains that the difference between a green and sustainable design is the degree to which the design helps maintain this ecological balance.

According to Cole (1999), there is a distinction between a "green" and "sustainable" building. The study suggests that green buildings improve the environmental performance of individual buildings. It also reflects that collective reduction in resource use and ecological loadings by the construction industry will be sufficient to fully address the environmental agenda (Cole, 1999). However, a "sustainable building" has environmental, social, and economic dimensions, and embraces all aspects of human activity (Cole, 1999). Similarly, Lützkendorf and Lorenz (2006) illustrated that a "green building" focuses on the environmental and health-related attributes of buildings, while a "sustainable building" looks at the inclusion of economic and social aspects that have resulted in a substantially widened scope. This fact is further strengthen by Shari and Soebarto (2012), and their research on applying green concept and sustainability concepts to a building.

Generally, sustainable development has three pillars, namely, environmental, social and economic (refer Section 2.2). Therefore, it can be concluded that when all these three are met in a building, it is a sustainable building. However, for the purpose of this study, the word "green building" is used and that term covers the essence of "sustainable development" with a focus on the triple bottom-line.

Green buildings always offer fruitful benefits to its occupants and for its developers. Studies (Bosch & Pearce, 2003; Nelms et al., 2005) suggest that green buildings are becoming more and more popular due to public awareness and perceived benefits. The key driver in going green, according to a survey conducted by McGraw-Hill Construction, United Technologies, and World Green Building Council (WGBC) (2013), is that now, green buildings are business imperatives around the world, and business drivers, such as client and market demand, are the key factors influencing the market. Further, according to the report, 76% of the respondents who were from the construction industry from over 60 countries, reported that green building lowered operating costs and more than one-third pointed to higher building values (38%), quality assurance (38%), and future-proofing assets (i.e. protecting against future demands—36%). These facts and figures clearly show that green buildings have become a trend in the construction industry.

After an extensive study Eichholtz, Kok, and Quigley (2010) illustrated that green buildings can offer many economic benefits. Initially, in this study, the researchers summarized that investments in energy efficiency at the time of construction and renovation could save current resources expended on energy, water, and waste disposal, as well as decrease other operation costs insured against future energy price increases. Further, improved corporate image through locating offices in green buildings can also attract premium prices and satisfied tenants, due to longer economic lives of the buildings that can provide more financial benefits by lower volatility in market value, as compared to conventional buildings (Eichholtz et al., 2010).

Usually, green office buildings achieve significantly higher rents estimated between 7.3% and 17.3%. Simultaneously, estimated occupancy levels are higher by approximately 10–18% when compared to other conventional office buildings (Wiley, Benefield, & Johnson, 2010). According to Green Building Council of Australia (2013a), in a green building, a minimal 2% upfront cost to support green design can result, on average, in life-cycle savings of 20% of total construction costs which is more than 10 times the initial investment, 15% net increase in perceived productivity for employees in offices and 25% improvement on test scores due to good lighting and ventilation in educational facilities.

In a report by the World Green Building Council (2013), it is stated that, due to the outside views provided by green buildings, there was 10–15% better mental function and memory, 6–12% faster call processing and 8.5% shorter hospital stays. Further, according to the report, daylight provided by green buildings compared to conventional buildings lead to increase of 5–14% higher test scores, 20–26% faster learning, 18% more productivity from workers and 15–40% increase in retail sales. Finally, the report concluded stating that there was a productivity increase by 23% from better lighting, 11% from better ventilation and 3% from individual temperature control in green buildings (World Green Building Council, 2013).

Considering all these facts and figures, it is evident that green buildings offer many benefits to society. However, with all these facts, a question commonly arises as to how to evaluate a green building, as opposed to a conventional building. If there is no distinction available, there will be buildings with poor environmental performance and positive communication about environmental performance leading to green-washing (Delmas & Burbano, 2011). While defining green buildings, there are many research studies with varying requirements as illustrated above. Therefore, to avoid accusations of green-washing within the industry, and to standardize the methods used to make buildings more environmentally friendly, green building rating tools were developed (Hoffman & Henn, 2008). Generally, these green building rating tools assess buildings and act as a solid yardstick in evaluating the building (Eichholtz et al., 2010).

The primary role of an environmental rating tool is to provide a comprehensive assessment of the environmental characteristics of a building using a common and verifiable set of criteria and targets for building owners and designers to achieve higher environmental standards (Cole & Larsson, 1999). Further, the assessment method reflects the significance of the concept of sustainable development in the context of building design and subsequent construction work on site (Ding, 2008).

There are many green building rating tools that have been developed by several different institutes and organizations in many countries, reflecting the requirements of each country. The first green building rating tool was launched in 1990, in the UK, named the Building Research Establishment Environmental Assessment Method (BREEAM) (Building Research Establishment Environment Assessment Method, 2015b). Later, the most discussed and widely used green building rating tool was launched by the United States Green Building Council (USGBC), named Leadership in Energy and Environmental Design (LEED) (United States Green Building Council, 2015a). The LEED building rating system is accepted in many countries in the world with 1.85 million square feet of construction space being certified every day (United States Green Building Council, 2015b). Further, there are many green building rating tools such as Green Star in Australia, Comprehensive Assessment System for Building Environmental Efficiency (CASBEE) in Japan, Green Mark in Singapore, and these were widely discussed and evaluated by many researchers (Crawley & Aho, 1999; Gowri, 2004; Haapio & Viitaniemi, 2008; Reed, Wilkinson, Bilos, & Schulte, 2011). Further, in order to identify the real essence of green buildings and their requirements, it is necessary to explore these green building rating tools in detail.

2.4.1 Green building rating tools

According to the World Green Building Council (2018), many countries have their own green building councils. These countries use their own green building rating systems or the most common and established rating tools to rate green buildings. The declared set of credit criteria identified by each rating tool (such as management, water efficiency, energy etc.) analyses how well or how poorly a building is performing, and is likely to perform (Cole, 2005). Therefore, the set of credit criteria identified by each green building rating tool has a critical impact on the evaluation of the building's performance. According to Lu, Geng, Liu, Cote, and Yu (2017), to pursue sustainable development, appropriate measurements are critical. In other words, developing the key credit criteria to evaluate green buildings is significant. Further, if the set credit criteria do not reflect the required performance of the building, the attempt to develop buildings in a more environmentally, socially and economically responsible way, would be in vain. Therefore, there are many green building rating tools that have been developed by many countries (Gowri, 2004; Haapio & Viitaniemi, 2008; Reed et al., 2011; Sinou & Kyvelou, 2006), all with the aim of reducing environmental impacts in both, construction and management phases of buildings (Asdrubali, Baldinelli, Bianchi, & Sambuco, 2015).

Initially, with the rising interest and demand from policy makers to achieve a sustainable society, there had been an increasing interest in environmental assessments of the built environment, focusing on energy use in buildings, the sick building syndrome, indoor climate, building materials containing hazardous substances, and/or many other aspects (Forsberg & Von Malmborg, 2004). Afterwards, separate environmental indicators were developed, such as energy and water efficiency measurements for the needs of relevant interest groups for building ratings (Haapio & Viitaniemi, 2008). However, these individual benchmarks serve to emphasize the need for a comprehensive rating tool to provide a thorough evaluation of building performance against a broad spectrum of environmental criteria (Ding, 2008). However, the first real attempt to establish comprehensive means of simultaneously assessing a broad range of environmental considerations in buildings was the BREEAM (Building Research Establishment Environment Assessment Method, 2015b; Crawley & Aho, 1999).

Usually, green building rating tools cover different phases of a building's life-cycle and take different environmental issues into account (Sachin & Jha, 2012; Waidyasekara, De Silva, & Rameezdeen, 2013),. Different tools are used to assess different types of buildings such as residential or office, influencing the choice of the environmental rating tool (Haapio & Viitaniemi, 2008). These tools can be used globally, nationally, and in some cases, locally as well. In developing green building rating tools, the environmental quality of buildings has to be taken into account during the design and construction phases, and also, the design performance criteria must be verified during the construction and commissioning phases (Kohler, 1999). However, to reduce the adverse impact on the environment, environmental assessment should be carried out from the initial design stage of the project (Crookes & de Wit, 2002).

According to Crawley and Aho (1999), from the construction and property sector's perspective, green building rating tools can be divided into two slightly different points of view: measuring the environmental impact of design, construction and property management activities (as services or industrial production processes), and the environmental impact of buildings (as products). Further, many of the existing green building rating tools can meet some of those needs (Crawley & Aho, 1999). The tools are divided into two categories, namely, assessment and rating tools.

Assessment tools provide quantitative performance indicators for design alternatives, while rating tools determine the performance level of a building in stars (Ding, 2008).

Reijnders and Van Roekel (1999) have made a rough division of rating tools into two classes, namely, qualitative tools based on scores and criteria, and quantitative tools using a physical life-cycle approach with quantitative input and output data. According to Forsberg and Von Malmborg (2004), qualitative methods are often based on audits of buildings, putting a score to each investigated parameter, resulting in one or several overall scores of a building. Further, in qualitative methods, some parameters that are investigated are quantitative, such as energy use, while others are entirely qualitative, based on specific criteria (Forsberg & Von Malmborg, 2004).

The stakeholders of any building have different interests over the building. For instance, a building owner may want his building to perform well from a financial point of view, whereas the occupants may be more concerned about IAQ, comfort, health, and safety issues (Ding, 2008). Therefore, using a single method to assess a building's environmental performance and to satisfy all needs of users is no easy task, but, an ideal green building rating tool will include all the requirements of the different parties involved in the development process (Ding, 2008). Although initial green building rating tools have focused on environmental improvements designed to produce "green" or "greener" buildings, now, there is a discussion on the need to bring sustainable development concerns into the tools, focusing on social and economic sustainability, as well (Todd, Crawley, Geissler, & Lindsey, 2001).

In green building rating tools "weighting" of criteria is one of the significant features because it dominates the overall performance score of the building being assessed (Lee, Chau, Yik, Burnett, & Tse, 2002). In general, weighting of criteria is used to incorporate regional differences into the green building rating tools. If there are no weightage factors given, all criteria are assumed to be of equal importance, and there is no order of importance for the criteria (Ding, 2008; Todd et al., 2001).

Over the past decade, it is evident that voluntary approaches and initiatives have been used increasingly in the construction industry by many parties to improve environmental performance, and help achieve sustainable development (United Nations Environment Programme, 2004). There are many green building rating tools which have been developed, launched and accepted by the public, and by now, have been widely used to assess the performance of green buildings worldwide (Asdrubali et al., 2015; Cole, 2005; Ding, 2008; Forsberg & Von Malmborg, 2004; Gowri, 2004; Haapio & Viitaniemi, 2008; Reed et al., 2011). Therefore, it is necessary to identify the green building rating tools that are widely used worldwide.

After an extensive study focusing on 71 green building councils worldwide Illankoon, Tam, Le, and Shen (2017), identified eight widely used green building rating tools representing the five main regions worldwide. These widely used rating tools are: LEED, BREEAM, Green Star from Australia, BEAM (Building Environmental Assessment Method) Plus from Hong Kong, CASBEE from Japan, Green Building Index (GBI) from Malaysia, and the Indian Green Building Council Rating (IGBC) from India (Illankoon et al., 2017). Each of these rating tools has different types of schemes, various ratings, and also various key criteria upon which the building is evaluated. Table 2.1 reports a summary of all these rating tools which are used widely, worldwide.

As illustrated in Table 2.1, each of these rating tools has different levels of certification and types of schemes. As an example, many LEED rating tools have been developed over time, since its inception, and have replaced the old versions. Now, the latest versions of these tools are identified as LEED Version 4 (refer Table 2.1). There are mainly five green building rating tools for different project types in LEED, namely, Building Design and Construction (BD+C), Interior Design and Construction (ID+C), Building Operations and Maintenance (O+M), Neighborhood Development (ND) and Homes. The number of points the project earns determines its level of LEED certification. There are four levels of certification in LEED, namely, certified, silver, gold, and platinum for each of these tools. Depending on the total points achieved by the building, the level of certification is calculated.

There are eight main credit criteria in the LEED rating tool. The key criteria are location and transport, sustainable sites, water efficiency, energy and atmosphere material and resources, indoor environment quality (IEQ), regional priority, and innovation (refer Table 2.1). Apart from these main eight credit criteria, there are three additional criteria for ND rating tools, namely, smart location and linkage, neighborhood pattern and design, and green infrastructure and buildings. There are also specific prerequisites that projects must satisfy, and a variety of credit points that projects can earn.

BREEAM was the first building assessment method in the world, since its inception in 1990 (Building Research Establishment Environment Assessment Method, 2015b). BREEAM sets the standard for best practices in sustainable building design, construction, and operation, and has become one of the most comprehensive and widely recognized measures of a building's environmental performance. BREEAM addresses wide-ranging environmental and sustainability issues, and enables developers, designers, and building managers to demonstrate the environmental credentials of their buildings to clients, planners, and other related parties (Building Research Establishment Environment Assessment Method, 2015b).

BREEAM is widely used in Europe and in many parts of the world, as well. Further, as of September 2015, BREEAM had certified over 425,000 projects and 1.9 million registered projects across 60 countries

TABLE 2.1 Summary of eight widely used green building rating tools.

	Green building rating tool							
Description	**LEED**	**Green Star**	**BREEAM**	**Green Mark**	**GBI**	**BEAM Plus**	**IGBC rating**	**CASBEE**
Parent organisation	United States Green Building Council (USGBC)	Green Building Council of Australia (GBCA)	Building Research Establishment (BRE)	Building Construction Authority (BCA)	Malaysian Institute for Architects (PAM architects)	Hong Kong Green Building Council	Indian Green Building Council (IGBC)	Japanese Sustainable Building Consortium Institute of Building Environment and Energy Conservation
Country	United States of America	Australia	United Kingdom	Singapore	Malaysia	Hong Kong	India	Japan
Year of origin	1994	2002	1990	2005	2008	1996	2003	2001
Type of ratings	• LEED Certified • LEED Silver • LEED Gold • LEED Platinum	• One Star—Not eligible for certification • Two Star—Not eligible for certification • Three Star—Not eligible for certification • Four Star—Green Star certified rating "Best Practice" • Five Star—Green Star certified rating "Australian Excellence" • Six star—Green Star certified rating "World Leader"	• Unclassified • Pass • Good • Very good • Excellent • Outstanding	• Green Mark Certified • Green Mark Gold • Green Mark Gold Plus • Green Mark Platinum	• Certified • Silver • Gold • Platinum	• Bronze • Silver • Gold • Platinum	• Certified—Good Practices • Silver—Best Practices • Gold—Outstanding Performance • Platinum—National Excellence • Super Platinum — Global Leadership	• Class C (poor) • Class B− • Class B+ • Class A • Class S (excellent)

Type of schemes available (latest in use)	LEED version 4 • Building Design and Construction (BD+C) • Interior Design and Construction (ID+C) • Building Operations and Maintenance (O+M) • Neighborhood Development (ND) • Homes	• Design and As built • Interiors • Communities • Performances	BREEAM International • BREEAM International New Construction (NC) • BREEAM International Refurbishment and Fit-Out • BREEAM In-Use International • BREEAM Communities Bespoke International	• Residential—new constructions • Residential—existing buildings • Non-residential—new constructions • Non-residential—existing buildings	• Non-Residential (NR) • Residential	• BEAM Plus for Existing and new buildings • BEAM Plus for Interiors	• IGBC New Buildings • IGBC Existing Buildings • IGBC Green Homes • IGBC Green Schools • IGBC Green Factory Buildings • IGBC Townships • IGBC Green SEZs • IGBC Green Landscapes • IGBC Green Mass Rapid Transit Systems	• CASBEE for pre-design (under-development) • CASBEE for new construction • CASBEE for existing buildings • CASBEE for renovations
Widely used scheme	Building Design and Construction (BD+C)	Design and As built	BREEAM International New Construction (NC)	Non-residential—new constructions	Non-Residential (NR)	BEAM Plus for Existing and new buildings	IGBC New Buildings	CASBEE for new construction

TABLE 2.1 Summary of eight widely used green building rating tools—cont'd

Description	Green building rating tool							
	LEED	Green Star	BREEAM	Green Mark	GBI	BEAM Plus	IGBC rating	CASBEE
Key criteria of widely used rating scheme	• Location and transport • Sustainable sites • Water efficiency • Energy and atmosphere • Material and resources • Indoor Environmental quality • Regional priority • Innovation	• Management • Indoor environment quality • Energy • Transport • Water • Material • Land use and ecology • Emissions • Innovation	• Management • Health and well-being • Energy • Transport • Water • Material • Waste • Land use and ecology • Pollution • Innovation	• Energy efficiency • Water efficiency • Environmental performances • Indoor Environmental quality • Other green features	• Energy efficiency • Indoor Environmental quality • Sustainable site planning and management • Material and resources • Water efficiency • Innovation	• Site aspects • Material aspects • Energy use • Water use • Indoor environmental quality • Innovation and additions	• Sustainable architecture and design • Site selection and planning • Water conservation • Building material and resources • Indoor environmental quality • Innovation and development	• Indoor environment (Q) • Quality of services (Q) • Outdoor environment (Q) • Energy (L_{BE}) • Resources and management (L_{BE}) • Off-site environment (L_{BE})

L_{BE}, built environment load; Q, built environment quality.

Source: Building Construction Authority. (2015). About BCA Green Mark scheme. Retrieved from http://www.bca.gov.sg/GreenMark/green_mark_buildings.html; Building Research Establishment Environment Assessment Method. (2015). What is BREEAM? Retrieved from http://www.breeam.org/; CASBEE. (2015). An overview of CASBEE. Retrieved from http://www.ibec.or.jp/CASBEE/english/overviewE.htm; Green Building Council of Australia. (2018). About GBCA. Retrieved from http://www.gbca.org.au/about/; Green Building Index. (2013). Green Building Index. Retrieved from Malaysia: http://www.greenbuildingindex.org/Resources/GBI%20Documents/GBI%20Explanatory%20Booklet%202013%2012.pdf; Hong Kong Green Building Council. (2015). About HKGBC. Retrieved from https://www.hkgbc.org.hk/eng/; Indian Green Building Council. (2015). Indian Green Building Council Retrieved from https://igbc.in/igbc/; United States Green Building Council. (2015). LEED. Retrieved from: http://www.usgbc.org/.

(Building Research Establishment Environment Assessment Method, 2015a). This rating tool is internationally recognized and there are eight country specific schemes available, all adapted to local conditions. The eight schemes are BREEAM international, UK, Germany, Netherlands, Spain, Norway, Sweden, and Austria. In other words, BREEAM international is tailor made to suit the conditions of other countries listed except the UK, and it is operated through a national scheme operator.

Green Mark scheme was launched in 2005 by Building Construction Authority Singapore in 2005 (Building Construction Authority, 2015). According to Building Construction Authority (2015), Green Mark certified buildings facilitate reduction in water and energy bills, reduce potential environmental impacts, improve IEQ for a healthy and productive workplace, and provide clear direction for continual improvement. Further, Building Construction Authority (BCA), Singapore, intended to promote sustainability in the built environment and raise environmental awareness among developers, designers, and builders when they start project conceptualization and design, as well as during construction. According to the Amendment Act 2008, all buildings with a gross floor area higher than 2000 m^2 must meet the Green Mark gold rating (Ahankoob, Morshedi, & Rad, 2013). From its inception in 2005, Green Mark has established its position in the Singapore market whereas, by 2014, a record number of 225 BCA Green Mark awards were awarded, including 78 platinum and 47 gold plus awards, by far the highest number of projects to have achieved high Green Mark ratings since the scheme was launched in 2005 (Building Construction Authority, 2014).

The Malaysian Institute for Architects (PAM architects) published the GBI in 2008, because the need for a localized green building rating tool became more evident, especially in light of increasing demands from building end-users for green-rated buildings that would not contribute to the destruction of the environment (Green Building Index, 2018). By now GBI has become the widely accepted and mostly used green building rating tool in Malaysia. As of September 2015, GBI has certified over 100 million square feet of green buildings overall (Green Building Index, 2018). Further, it is necessary to note that in Malaysia, for any person who incurs additional expenses to register for green building rating through GBI certification for a building used for his business qualifies for tax exemption (Green Building Index, 2013). A stamp duty exemption is also provided on instruments of transfer of ownership of buildings and residential properties acquired from property developers awarded with a GBI certificate (Green Building Index, 2013). This clearly illustrates the significance that GBI has gained over these years in the Malaysian market.

BEAM Plus is a green building rating tool established by the Hong Kong Green Building Council. Initially, this was identified as the Hong

Kong Building Environmental Assessment Method (HK-BEAM) scheme which was established in 1996, largely based on the UK Building Research Establishment's BREEAM. There was a significant upgrade to the previous BEAM documents in 2004 (Hong Kong Green Building Council, 2015). In 2009, in response to the critical global environmental issue, BEAM was developed further to meet higher expectations of the public and the community at large. BEAM Plus is now one of the prerequisites for the granting of gross floor area concessions for certain green and amenity features in development projects (Hong Kong Green Building Council, 2015). Since the early stages, BEAM Plus has gained wider recognition in the Hong Kong green building market and has certified many projects in both, new and existing building and interior schemes.

IGBC rating is the green building rating tool developed and used in India. With a modest beginning of 20,000 ft^2, which was the green built-up area in the country in the year 2003, it has grown to more than 3356 green buildings projects in as on October 2015. More are coming up, with a footprint of over 3.127 billion square feet having been registered with the IGBC, out of which 698 green building projects are certified and fully functional in India (Indian Green Building Council, 2015). Further, according to the Indian Green Building Council (2015), India is the second largest country in the world with the largest green building footprint. These figures clearly show the rapid recognition that the IGBC rating has gained in India over the past decade. This rating system is based on the five elements of nature and is a perfect blend of ancient architectural practices and modern technological innovations. The ratings system is applicable to all five climatic zones in the country (Indian Green Building Council, 2015).

CASBEE is a green building rating tool which was developed by the Japanese Sustainable Building Consortium and Institute of Building Environment and Energy Conservation in 2001. This rating tool is specifically designed considering the specific climatic and environmental conditions in Japan (Suzer, 2015). Therefore, compared to other green building rating tools, CASBEE is different in structure. The four target fields of CASBEE, namely, energy and resource efficiency, and local and indoor environments, are arranged and categorized into environmental quality and built environmental load. Finally, based on the values obtained for each criterion of built environmental quality and built environmental load, the building environment efficiency (BEE) is calculated. In CASBEE, each requirement is evaluated on a scale of 5, and there is no direct point allocation like in other green building rating tools.

In the Australian context, Green Star is the widely used green building rating tool. This study follows the Green Star rating for developing the proposed model.

2.4.2 Green Star rating tool

The Green Building Council of Australia (GBCA) was launched in 2002 as a national not-for-profit organization focusing on the development of the sustainable property industry in Australia (Green Building Council of Australia, 2018a). In 2003, GBCA launched its green building rating system as "Green Star." Australia being the largest single contributor of greenhouse gases, generating 40% of waste, Green Star is aimed at improving environmental efficiency in buildings, while also boosting productivity, creating jobs, and improving the health and well-being of communities (Green Building Council of Australia, 2013b).

Initially, GBCA had Green legacy tools, and by the end of 2015, all these tools were superseded by a new set of green building rating tools. These tools include four green building rating tools, namely, "Green Star—Design and As Built," "Green Star—Interiors," "Green Star—Communities," and "Green Star—Performance." Each of these green building rating tools is applied for different sectors in the built environment. Further, in Green Star there is a scale of 6 stars, starting from one star for minimum practice, and six stars for world leadership. Further, Green Star—Design and As-Built, Interiors, and Communities projects can achieve Green Star certifications of 4 to 6 Stars. Buildings assessed using the Green Star—Performance rating tool can achieve a Green Star rating from 1 to 6 Stars (Green Building Council of Australia, 2013b).

Both, in new green building rating tools and legacy tools, there are nine environmental categories, namely, management (M), IEQ, energy (E), transport (T), water (W), materials (Mat), emissions (Em), land use and ecology (L) and innovation (I) (refer Table 2.1). However, in Green Star Performance tool, there are six different categories on which the community is evaluated. These categories are governance, design, liveability, economic prosperity, environment, and innovation. These categories are given credits, and the final score is the sum of all the credit points. Final certification is awarded based on the final score. Table 2.2 reports the scores and outcomes for Green Star certification.

There is an exception for green building rating for the Green Star—Performance tool for certification, whereas green certification is available for one, two and three star ratings as well (Green Building Council of Australia, 2013c).

According to the Green Star rating tool, there are five steps to be followed in the certification process. Initially, the project must be registered online, and then, a documentation process follows to ensure that the project meets the relevant benchmarks. After submission, the documents are reviewed by an independent panel of sustainable development experts and an overall score is assigned. Finally, Green Star certified rating is

TABLE 2.2 Green Star green building rating tools—scores and outcome.

Overall score	Rating	Outcome	Remarks
Less than 10	Zero star	Assessed	Not eligible for certification
10–19	One star	Minimum practice	Not eligible for certification
20–29	Two star	Average practice	Not eligible for certification
30–44	Three star	Good practice	Not eligible for certification
45–59	Four star	Best practice	Eligible for certification
60–74	Five star	Australian excellence	Eligible for certification
75+	Six star	World leader	Eligible for certification

Source: Green Building Council of Australia. (2015). Green Star—Design and As Built version 1.1. Retrieved from https://www.gbca.org.au/green-star/green-star-design-as-built/the-rating-tool/.

awarded as a third-party verification of a project's sustainability (Green Building Council of Australia, 2013b).

Due to spatial differentiation requirements, a project must be distinct and separate. Project components are not eligible for certification (Green Building Council of Australia, 2018b). Apart from this, all green building rating tools have conditional requirements, such as maximum greenhouse gas emissions and protection of land with high ecological value. These different credits and their requirements are discussed in detail in the later chapters.

In the past, many projects were certified according to these Green Star legacy rating tools. Fig. 2.1 reports the Green Star projects certified from 2003 to October 2012 using green legacy tools.

According to Fig. 2.1, 58% of the buildings that were certified were office buildings. Further, 22% were office interiors. The educational buildings constituted 11% of the total. Therefore, according to this, there was a clear demand for the office buildings to be green certified. With reference to Fig. 2.1, if new tools are used, except for office interiors, all the others would be assessed based on the Green Star Design and As-Built rating tool, which would constitute approximately 62% of the certified projects. Therefore, it can be predicted that in comparison to the other three, Green Star Design and As-Built would be mostly used in practice in the future.

Irrespective of the building type (excluding buildings classified under Building Code of Australia Class 7a and Class 10), according to Green Star rating tool, buildings can be registered and certified under Green Star Design and As-Built (Green Building Council of Australia, 2015). However, in terms of the type of building, a majority of the certified projects are commercial office buildings (refer Fig. 2.1). Therefore, considering these two facts, this research focuses on Green Star Design and As-Built version 1.1 rating tools, for commercial office buildings.

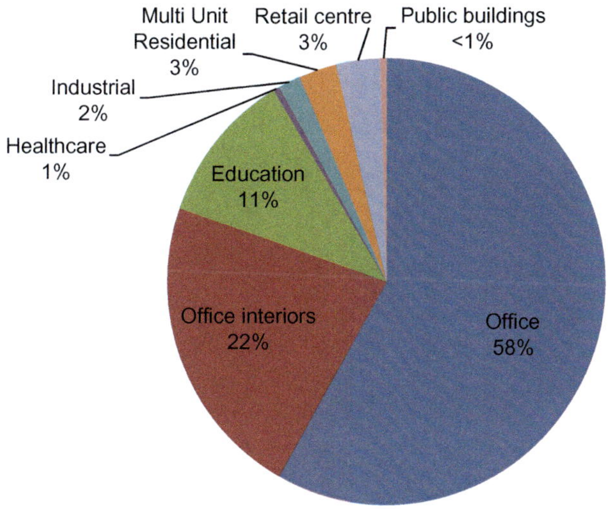

FIG. 2.1 Green Star certified projects from 2003 to October 2012. Data Source: Green Building Council of Australia. (2013). The value of Green Star—A decade of environmental benefits. Retrieved from https://www.gbca.org.au/uploads/194/34754/The_Value_of_Green_Star_A_Decade_of_Environmental_Benefits.pdf.

2.5 Triple bottom-line in green building certification

According to Section 2.4, green buildings should consider fulfilling the triple bottom-line requirements, that is, it should fulfill the requirements for environmental, social and economic sustainability. However, after a thorough investigation of different credits and credit points of all the eight green building rating tools (refer Table 2.1), Illankoon, Tam, and Le (2016) illustrated that although the three pillars of sustainability were considered equally important, economic sustainability was not greatly considered in green building rating tools. This fact is further strengthened by Illankoon et al. (2017) after evaluating the credits points and key criteria of various green building rating tools used worldwide. Further, the economic pillar of sustainability is overlooked in existing green building rating tools, and therefore, Illankoon et al. (2017); Illankoon, Tam, Le, and Shen (2016) suggested that to provide a better evaluation, criteria focusing on economic sustainability must be taken into consideration in developing green building rating tools.

In green building certification, the triple bottom-line is represented by various criteria in green building rating tools. Therefore, Illankoon et al. (2017), thoroughly evaluated the eight green building rating tools (refer Table 2.1) and classified the credits under key criteria, namely, site, energy, water, IEQ, material, waste and pollution, and management. Fig. 2.2 below reports the analysis.

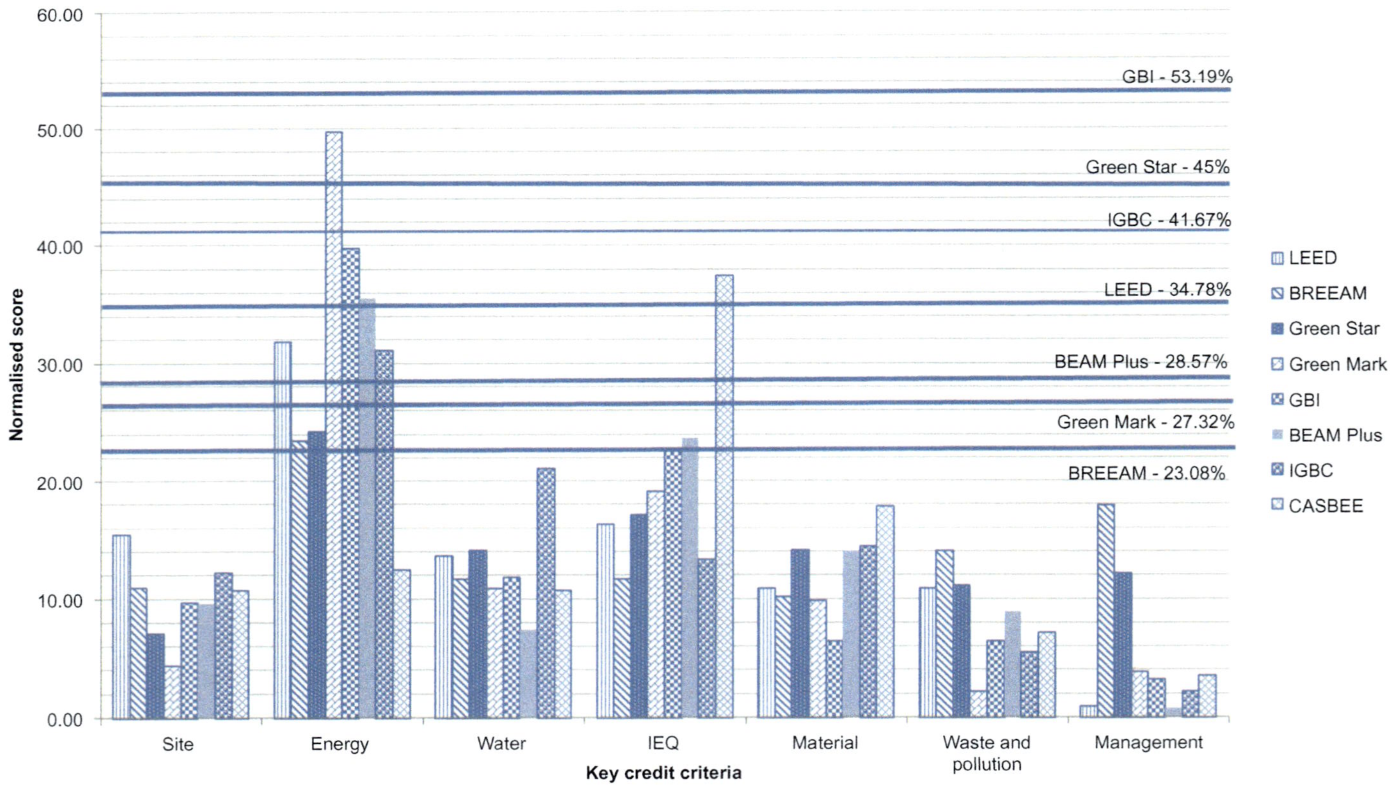

FIG. 2.2 Key criteria and certification levels of eight selected green building rating tools. Source: Illankoon, I.M.C.S., Tam, V.W., Le, K.N., & Shen, L. (2017). Key credit criteria among international green building rating tools. *Journal of Cleaner Production, 164*, 209–220 (p. 216).

There are certain minimum scores in order to obtain the lowest certification in each green building rating tool (refer Table 2.1). In Fig. 2.2, the horizontal lines show the normalized scores that are required for minimum certification of each green building rating tool. The highest required score is from GBI, and the lowest is BREEAM (refer Fig. 2.2). According to Fig. 2.2, in LEED, BEAM Plus, and Green Mark, the minimum normalized score required for certification is significantly lower than the normalized score allocated to the energy criterion. However, green buildings cater to all the triple bottom-line aspects. Therefore, concentrating on one criterion does not cater to the requirements of developing a green building. As an example, if a building obtains all the credit points in one criterion alone, it does not mean that it operates as a green building. However, in this situation, where the minimum normalized score required for certification is lesser than the normalized score of an energy criterion, there is a possibility that a building might be given a green certification, although it does not focus on other green requirements such as water, IEQ, and so on.

The minimum credit requirement in BREEAM is slightly lower than that of the energy criterion (refer Fig. 2.2). Therefore, LEED, BEAM Plus and Green Mark certification can be obtained by focusing only on the energy criterion. However, in Green Star, IGBC and GBI, the minimum credit point requirement is higher than the key credit criteria, with the highest normalized score (refer Fig. 2.2). Even though the minimum requirement of total credits is higher than one specific credit, these buildings can obtain certification ignoring one or more key criteria. Therefore, there is a possibility that these green certifications can be obtained without considering a holistic approach to the triple bottom-line.

2.6 Summary

Sustainable development is a lively field for research today. It has mainly gained recognition and opened the eyes of the public since the Brundtland Report in 1987. Sustainable development has evolved into many diverse areas, sustainable construction being one, which has also gained attention in research since inception. There are many concepts underlying sustainable development, but the most widely discussed concept is the triple bottom-line, indicating that sustainable development focuses on social, economic, and environmental sustainability.

In general terms, social sustainability focuses on the social well-being of human beings. Economic sustainability refers to the financial costs and benefits. Environmental sustainability focuses on environmental issues and the impact of human activities on the environment. However, there are criticisms of these three aspects of sustainability, and it has been discussed that sustainable development should not be confined to these three aspects, but also focus on technical, political, legal, and other criteria, as well. According to the literature, sustainable development is not static. It is an evolving process. Sustainable development should re-shape itself to cater to the requirements of human needs, but while still retaining its core essence.

As far as the human needs are considered, construction plays a major role in society. After considering the excessive negative impacts on the environment by conventional construction, sustainable construction is considered a greater priority. Initially, sustainable construction was all about providing healthy buildings to occupants with a minimal impact on the environment. However, in the current scenario, it has broadened its scope. Currently, sustainable construction should look into the social, economic, and environmental aspects of construction. Further, sustainable construction is also a process from cradle to grave, lasting throughout the entire life-cycle from the inception to demolition. It adopts concepts such as resource reuse, minimum resource usage, use of renewable resources, and providing a healthy environment throughout the life-cycle. Finally, it draws the reader's attention to green buildings.

Green buildings can be identified as environmental friendly buildings which provide healthy environments for its occupants, fulfilling the triple bottom-line sustainability. These buildings operate in an efficient way in terms of water, energy, and material usage, and many other things. Due to the numerous benefits it provides, it has gained wide recognition. Usually, green buildings are considered to have lower operation costs and higher productivity rates.

In the literature, there are two distinctions between green and sustainable buildings. Some argue that both are more or less the same, but others argue that green buildings focus only on environmental aspects, whereas sustainable buildings focus on the threefold aspects of social, economic, and environmental development. For the purpose of this study, green buildings are responsible for all social, economic, and environmental considerations.

In order to reduce and regulate adverse environmental impacts, many green building rating tools were developed as a yardstick. Initially, the tools were developed specifically focusing on one aspect, such as energy or water efficiency. The tools were used to identify the efficiency of buildings in terms of one or two criteria. However, with the passing of time, in 1990, BREEAM was developed to evaluate buildings in a more detailed and broader manner, considering all the necessary criteria at once. Since then, many green building rating tools have been developed based on that principle.

At present, there are many green building councils around the world which promote the construction of green buildings. In order to represent the main regions in the world, this study identified eight green building rating tools for evaluation. These green building rating tools were BREEAM, LEED, Green Star, GBI, Green Mark, BEAM Plus, IGBC rating, and CASBEE. Further, these green building rating tools were developed by various institutions in different countries. Each of these rating tools has different key criteria and certification levels which have been tabulated in this chapter. Finally, this study compared the certification levels with the credits points allocated for each criterion in green building rating tools. Based on this analysis, there is a possibility for green building rating tools to obtain certification, even though certain criteria have been completely eliminated.

References

Abidin, N. Z. (2010). Investigating the awareness and application of sustainable construction concept by Malaysian developers. *Habitat International, 34*(4), 421–426. https://doi.org/10.1016/j.habitatint.2009.11.011.

Ahankoob, A., Morshedi, R., & Rad, K. G. (2013). A comprehensive comparison between LEED and BCA Green Mark as green building assessment tool. *The International Journal of Engineering and Sciences, 2*(7), 31–38.

Asdrubali, F., Baldinelli, G., Bianchi, F., & Sambuco, S. (2015). A comparison between environmental sustainability rating systems LEED and ITACA for residential buildings. *Building and Environment, 86*, 98–108. https://doi.org/10.1016/j.buildenv.2015.01.001.

ASHRAE. (2006). Green/sustainable high-performance design. In ASHRAE (Ed.), *The ASHRAE greenguide: The design, construction, and operation of sustainable buildings* (2nd ed., pp. 3–16). Burlington: Butterworth-Heinemann.

Balkema, A. J., Preisig, H. A., Otterpohl, R., & Lambert, F. J. (2002). Indicators for the sustainability assessment of wastewater treatment systems. *Urban Water, 4*(2), 153–161. https://doi.org/10.1016/S1462-0758(02)00014-6.

Baloi, D. (2003). Sustainable construction: Challenges and opportunities. In *Paper presented at the 19th annual ARCOM conference*.

Bosch, S. J., & Pearce, A. R. (2003). Sustainability in public facilities: Analysis of guidance documents. *Journal of Performance of Constructed Facilities, 17*(1), 9–18. https://doi.org/10.1061/(ASCE)0887-3828(2003)17:1(9).

Brundtland, G. H. (1987). *Report of the world commission on environment and development: Our common future (A/42/427)*. Retrieved from Oslo: http://www.un-documents.net/our-common-future.pdf.

Building Construction Authority. (2014). *More building projects achieving higher BCA Green Mark standards [Press release]*. Retrieved from http://www.news.gov.sg/public/sgpc/en/media_releases/agencies/bca/press_release/P-20140429-1/AttachmentPar/0/file/BCA%20Media%20Release%20-%20Green%20Mark%20Awards%202014.pdf.

Building Construction Authority. (2015). *About BCA Green Mark scheme*. Retrieved from http://www.bca.gov.sg/GreenMark/green_mark_buildings.html.

Building Research Establishment Environment Assessment Method. (2015a). *Explore BREEAM*. Retrieved from http://www.breeam.com/projects/explore/index.jsp.

Building Research Establishment Environment Assessment Method. (2015b). *What is BREEAM?*. Retrieved from http://www.breeam.org/.

Carew, A., & Mitchell, C. (2008). Teaching sustainability as a contested concept: Capitalizing on variation in engineering educators' conceptions of environmental, social and economic sustainability. *Journal of Cleaner Production, 16*(1), 105–115. https://doi.org/10.1016/j.jclepro.2006.11.004.

Cassidy, R., Wright, G., & Flynn, L. (2003). *White paper on sustainability: A report of the green building movement.* Building design and construction Clearwater, FL: Reed Business Information.

Chan, E. H. W., Qian, Q. K., & Lam, P. T. I. (2009). The market for green building in developed Asian cities—The perspectives of building designers. *Energy Policy, 37*(8), 3061–3070. https://doi.org/10.1016/j.enpol.2009.03.057.

Cole, R. J. (1999). Building environmental assessment methods: Clarifying intentions. *Building Research and Information, 27*(4–5), 230–246. https://doi.org/10.1080/096132199369354.

Cole, R. J. (2005). Building environmental assessment methods: Redefining intentions and roles. *Building Research and Information, 33*(5), 455–467. https://doi.org/10.1080/09613210500219063.

Cole, R., & Larsson, N. K. (1999). GBC'98 and GBTool: Background. *Building Research and Information, 27*(4-5), 221–229.

Crawley, D., & Aho, I. (1999). Building environmental assessment methods: Applications and development trends. *Building Research and Information, 27*(4-5), 300–308. https://doi.org/10.1080/096132199369417.

Crookes, D., & de Wit, M. (2002). Environmental economic valuation and its application in environmental assessment: An evaluation of the status quo with reference to South Africa. *Impact Assessment and Project Appraisal, 20*(2), 127–134. https://doi.org/10.3152/147154602781766753.

Curran, M. A. (2009). Wrapping our brains around sustainability. *Sustainability, 1*(1), 5–13.

Delmas, M. A., & Burbano, V. C. (2011). The drivers of greenwashing. *California Management Review, 54,* 1–64.

Ding, G. K. C. (2008). Sustainable construction—The role of environmental assessment tools. *Journal of Environmental Management, 86*(3), 451–464. https://doi.org/10.1016/j.jenvman.2006.12.025.

Dovers, S. R., & Handmer, J. W. (1992). Uncertainty, sustainability and change. *Global Environmental Change, 2*(4), 262–276.

DuPlessis, C. (2007). A strategic framework for sustainable construction in developing countries. *Construction Management and Economics, 25*(1), 67–76. https://doi.org/10.1080/01446190600601313.

Eichholtz, P., Kok, N., & Quigley, J. M. (2010). Doing well by doing good? Green office buildings. *The American Economic Review,* 2492–2509. https://doi.org/10.1257/aer. 100.5.2492.

Elkington, J. (1994). Towards the sustainable corporation: Win-win-win business strategies for sustainable development. *California Management Review, 36*(2), 90–100.

Forsberg, A., & Von Malmborg, F. (2004). Tools for environmental assessment of the built environment. *Building and Environment, 39*(2), 223–228. https://doi.org/10.1016/j.buildenv.2003.09.004.

Foxon, T. J., McIlkenny, G., Gilmour, D., Oltean-Dumbrava, C., Souter, N., Ashley, R., ... Moir, J. (2002). Sustainability criteria for decision support in the UK water industry. *Journal of Environmental Planning and Management, 45*(2), 285–301. https://doi.org/10.1080/09640560220116341.

Giddings, B., Hopwood, B., & O'brien, G. (2002). Environment, economy and society: Fitting them together into sustainable development. *Sustainable Development, 10*(4), 187–196.

Gladwin, T. N., Kennelly, J. J., & Krause, T.-S. (1995). Shifting paradigms for sustainable development: Implications for management theory and research. *Academy of Management Review, 20*(4), 874–907.

Global Reporting Initiative. (2011). *Sustainability reporting guidelines.* Retrieved from https://www.globalreporting.org/resourcelibrary/G3.1-Guidelines-Incl-Technical-Protocol.pdf.

Gowri, K. (2004). Green building rating systems: An overview. *ASHRAE Journal, 46*(11), 56.

Green Building Council of Australia. (2013a). *The business case for green buildings.* Retrieved from https://www.gbca.org.au/uploads/63/34623/Evolution_2013_Business_Case_for_Green_Building.pdf.

Green Building Council of Australia. (2013b). *Introducing Green Star*. Retrieved from https://www.gbca.org.au/resources/gbca-publications/introducing-green-star/.

Green Building Council of Australia. (2013c). *The value of Green Star—A decade of environmental benefits*. Retrieved from https://www.gbca.org.au/uploads/194/34754/The_Value_of_Green_Star_A_Decade_of_Environmental_Benefits.pdf.

Green Building Council of Australia. (2015). *Green Star—Design and As Built version 1.1*. Retrieved from https://www.gbca.org.au/green-star/green-star-design-as-built/the-rating-tool/.

Green Building Council of Australia. (2018a). *About GBCA*. Retrieved from http://www.gbca.org.au/about/.

Green Building Council of Australia. (2018b). *Green Star rating tools*. Retrieved from https://www.gbca.org.au/green-star/rating-tools/.

Green Building Index. (2013). *Green Building Index*. Retrieved from Malaysia: http://www.greenbuildingindex.org/Resources/GBI%20Documents/GBI%20Explanatory%20Booklet%202013%2012.pdf.

Green Building Index. (2018). *How GBI works*. Retrieved from http://www.greenbuildingindex.org/how-GBI-works.html.

Grierson, D. (2009). Towards a sustainable built environment. *CIC Start Online Innovation Rev, 1*, 70–77.

Haapio, A., & Viitaniemi, P. (2008). A critical review of building environmental assessment tools. *Environmental Impact Assessment Review, 28*(7), 469–482. https://doi.org/10.1016/j.eiar.2008.01.002.

Hill, R. C., & Bowen, P. A. (1997). Sustainable construction: Principles and a framework for attainment. *Construction Management and Economics, 15*(3), 223–239. https://doi.org/10.1080/014461997372971.

Hoffman, A. J., & Henn, R. (2008). Overcoming the social and psychological barriers to green building. *Organization & Environment, 21*(4), 390–419. https://doi.org/10.1177/1086026608326129.

Hong Kong Green Building Council. (2015). *About HKGBC*. Retrieved from https://www.hkgbc.org.hk/eng/.

Illankoon, I.M.C.S., Tam, V. W. Y., & Le, K. N. (2016). Environmental, Economic, and Social Parameters in International Green Building Rating Tools. *Journal of Professional Issues in Engineering Education and Practice, 143*(2), 457–467.

Illankoon, I.M.C.S., Tam, V. W. Y., Le, K. N., & Shen, L. (2016). Environmental, social and economical parameters of sustainability in green building rating tools worldwide. In *Paper presented at the Australasian Universities Building Education Association Conference (AUBEA), Cairns, Australia.*

Illankoon, I.M.C.S., Tam, V. W., Le, K. N., & Shen, L. (2017). Key credit criteria among international green building rating tools. *Journal of Cleaner Production, 164*, 209–220.

Indian Green Building Council. (2015). *Indian Green Building Council*. Retrieved from https://igbc.in/igbc/.

International Council for Research and Innovation in Building and Construction. (1999). *Agenda 21 on sustainable construction (237)*. Retrieved from http://cic.vtt.fi/eco/cibw82/A21text.pdf.

Kats, G., Alevantis, L., Berman, A., Mills, E., & Perlman, J. (2003). *The costs and financial benefits of green buildings*. A report to California's sustainable building task force. USA.

Kibert, C. J. (1994). Establishing principles and a model for sustainable construction. In: *Paper presented at the proceedings of the first international conference on sustainable construction*.

Kibert, C. J. (2012). *Sustainable construction: Green building design and delivery*. John Wiley & Sons.

Kohler, N. (1999). The relevance of green building challenge: An observer's perspective. *Building Research and Information, 27*(4–5), 309–320.

Kuhlman, T., & Farrington, J. (2010). What is sustainability? *Sustainability, 2*(11), 3436–3448.

Lavy, S., & Fernández-Solis, J. L. (2009). LEED accredited professionals' perceptions affecting credit point adoption. *Facilities, 27*(13/14), 531–548. https://doi.org/10.1108/02632770910996360.

Lee, W., Chau, C., Yik, F., Burnett, J., & Tse, M. (2002). On the study of the credit-weighting scale in a building environmental assessment scheme. *Building and Environment, 37*(12), 1385–1396. https://doi.org/10.1016/S0360-1323(02)00006-9.

Lu, Y., Geng, Y., Liu, Z., Cote, R., & Yu, X. (2017). Measuring sustainability at the community level: An overview of China's indicator system on National Demonstration Sustainable Communities. *Journal of Cleaner Production, 143*, 326–335. https://doi.org/10.1016/j.jclepro.2016.12.105.

Lützkendorf, T., & Lorenz, D. (2006). Using an integrated performance approach in building assessment tools. *Building Research and Information, 34*(4), 334–356. https://doi.org/10.1080/09613210600672914.

McGraw-Hill Construction, United Technologies & World Green Building Council (WGBC). (2013). *World green building trends.* Retrieved from Bedford, MA 01730: http://www.worldgbc.org/infohub/global-green-building-trends/.

Munasinghe, M. (2004). *Sustainomics: A trans-disciplinary framework for making development more sustainable.* Colombo, Sri Lanka: International Society for Ecological Economics.

Nelms, C., Russell, A. D., & Lence, B. J. (2005). Assessing the performance of sustainable technologies for building projects. *Canadian Journal of Civil Engineering, 32*(1), 114–128.

Olawumi, T. O., & Chan, D. W. M. (2018). A scientometric review of global research on sustainability and sustainable development. *Journal of Cleaner Production, 183*, 231–250. https://doi.org/10.1016/j.jclepro.2018.02.162.

Ortiz, O., Castells, F., & Sonnemann, G. (2009). Sustainability in the construction industry: A review of recent developments based on LCA. *Construction and Building Materials, 23*(1), 28–39.

Parkin, S. (2000). Sustainable development: The concept and the practical challenge. In *Paper presented at the Proceedings of the ICE-Civil Engineering*.

Pawłowski, A. (2008). How many dimensions does sustainable development have? *Sustainable Development, 16*(2), 81–90. https://doi.org/10.1002/sd.339.

Pirages, D. (1994). Sustainability as an evolving process. *Futures, 26*(2), 197–205. https://doi.org/10.1016/0016-3287(94)90109-0.

Reed, R., Wilkinson, S., Bilos, A., & Schulte, K.-W. (2011). A comparison of international sustainable building tools–An update. In *Paper presented at the 17th annual pacific rim real estate society conference, Gold Coast.*

Reijnders, L., & Van Roekel, A. (1999). Comprehensiveness and adequacy of tools for the environmental improvement of buildings. *Journal of Cleaner Production, 7*(3), 221–225.

Ross, A. (2009). Modern interpretations of sustainable development. *Journal of Law and Society, 36*(1), 32–54.

Sachin, V. G., & Jha, K. N. (2012). Comparative study of rating systems for green building in developing and developed countries. In: *Paper presented at the third international conference on construction in developing countries (ICCIDC–III) "advancing civil, architectural and construction engineering & management", Bangkok, Thailand.*

Said, H., & Berger, L. (2013). Future trends of sustainability design and analysis in construction industry and academia. *Practice Periodical on Structural Design and Construction*, https://doi.org/10.1061/(ASCE)SC.1943-5576.0000181.

Sartori, S., Latrônico, F., & Campos, L. (2014). Sustainability and sustainable development: A taxonomy in the field of literature. *Ambiente & Sociedade, 17*(1), 1–22.

Shah, S. (2008). *Sustainable practice for the facilities manager.* John Wiley & Sons.

Shari, Z., & Soebarto, V. (2012). Green vs. Sustainability performance assessment: A case study of an office building in putrajaya, Malaysia. In *Paper presented at the proceedings—28th international PLEA conference on sustainable architecture + urban design: Opportunities, limits and needs—towards an environmentally responsible architecture, PLEA 2012.*

Sinou, M., & Kyvelou, S. (2006). Present and future of building performance assessment tools. *Management of Environmental Quality: An International Journal, 17*(5), 570–586. https://doi.org/10.1108/14777830610684530.

Solow, R. M. (1992). *An almost practical step toward sustainability*. Resources for the Future.

Sutton, P. (2004). What is sustainability. *Eingana, The Journal of the Victorian Association for Environmental Education*. 1–7.

Suzer, O. (2015). A comparative review of environmental concern prioritization: LEED vs other major certification systems. *Journal of Environmental Management, 154*, 266–283. https://doi.org/10.1016/j.jenvman.2015.02.029.

Svensson, G., & Wagner, B. (2015). Implementing and managing economic, social and environmental efforts of business sustainability: Propositions for measurement and structural models. *Management of Environmental Quality: An International Journal, 26*(2), 195–213. https://doi.org/10.1108/SCM-02-2014-0061.

Todd, J. A., Crawley, D., Geissler, S., & Lindsey, G. (2001). Comparative assessment of environmental performance tools and the role of the green building challenge. *Building Research and Information, 29*(5), 324–335. https://doi.org/10.1080/09613210110064268.

United Nations Commssion on Sustainable Development. (2001). *Indicators of sustainable development: Guidelines and methodologies (E/CN.17/2001/19)*. Retrieved from http://www.un.org/ga/search/view_doc.asp?symbol=E/CN.17/2001/19%20(SUPP)&Lang=E.

United Nations Environment Programme. (2004). Sustainable consumption and production (ISSN 0378-9993). In: Retrieved from 8th international high-level seminar on sustainable consumption and production, Monterrey, Mexico.

United States Environment Protection Agency. (2014). *Green building*. Retrieved from http://archive.epa.gov/greenbuilding/web/html/about.html.

United States Green Building Council. (2007). *Making the business case for high performance green buildings*. Retrieved from 1800 Massachusetts Ave, NW Suite 300 Washington, DC 20036 http://www.usgbc.org/Docs/Member_Resource_Docs/makingthebusinesscase.pdf.

United States Green Building Council. (2015a). *LEED*. Retrieved from: http://www.usgbc.org/.

United States Green Building Council. (2015b). *LEED digital marketing brochure*. Retrieved from http://leed.usgbc.org/.

Vanegas, J. A., DuBose, J. R., & Pearce, A. R. (1996). Sustainable technologies for the building construction industry. In *Paper presented at the proceedings, symposium on design for the global environment, Atlanta, GA*.

Vos, R. O. (2007). Defining sustainability: A conceptual orientation. *Journal of Chemical Technology and Biotechnology, 82*(4), 334–339. https://doi.org/10.1002/jctb.1675.

Waidyasekara, K., De Silva, M., & Rameezdeen, R. (2013). Comparative study of green building rating systems: In terms of water efficiency and conservation. In *Paper presented at the second world construction symposium 2013*.Colombo, Sri Lanka: Socio Economic Sustainability in Construction.

Welsford, J., See, Y. H., & Erkki, N. J. (2014). Sustainability—India in global context. *Journal of Indian Pulp and Paper Technical Association, 26*(1), 47–52.

Wilbanks, J. T., & Wilbanks, T. J. (2010). Science, open communication and sustainable development. *Sustainability, 2*(4), 993–1015. https://doi.org/10.3390/su2040993.

Wiley, J. A., Benefield, J. D., & Johnson, K. H. (2010). Green design and the market for commercial office space. *The Journal of Real Estate Finance and Economics, 41*(2), 228–243. https://doi.org/10.1007/s11146-008-9142-2.

Woolley, T., Kimmins, S., Harrison, P., & Harrison, B. (2005). *Green building handbook. Vol. 1*. New York: E & FN Spon.

World Green Building Council. (2013). *The business case for green buildings*. Retrieved from http://www.worldgbc.org/activities/business-case/.

World Green Building Council. (2018). Retrieved from http://www.worldgbc.org/index.php?cID=1.

World Watch Institute. (2015). *Vision for a sustainable world*. Retrieved from http://www.worldwatch.org/.

World Wildlife Fund. (2015). *Green building design*. Retrieved from http://www.wwf.org.au/get_involved/change_the_way_you_live/green_building_design/.

Wyatt, D. P. (1994). Recycling and serviceability: The twin approach to securing sustainable construction. In *Paper presented at the proceedings of first international conference of CIB TG*.

Ye, K., Zhu, W., Shan, Y., & Li, S. (2015). Effects of market competition on the sustainability performance of the construction industry: China case. *Journal of Construction Engineering and Management*, *141*(9), 04015025.

Young, J. W. (1997). A framework for the ultimate environmental index–putting atmospheric change into context with sustainability. *Environmental Monitoring and Assessment*, *46*(1), 135–149.

CHAPTER

3

Cost considerations of green buildings

3.1 Introduction

This chapter focuses on the cost considerations of green buildings. First, this chapter discusses the cost premium of green buildings. Afterwards, this chapter focuses on the life-cycle cost for green building implementation. The final section of this chapter discusses the various life-cycle cost models developed for green buildings. These different cost models are evaluated, and the limitations of these life-cycle cost models are discussed.

3.2 Initial cost premium of green buildings

Green buildings have become a buzzword in this era, yet there are many misconceptions and myths attached to this concept. "Cost" is one of the main topics discussed whenever the topic surfaces. There is a basic idea that the cost of green buildings is higher compared to conventional buildings, and there are many other counterarguments, too. Therefore, it is necessary to initially identify the actual scenario of the cost of green buildings.

This scenario of cost is necessary for informed decision-making on a number of levels. The private entrepreneur needs to know whether green construction involves added costs and, if so, what the payback period is. The public needs to be made aware of the economic benefits of green buildings for the national economy as well as for their users (Gabay, Meir, Schwartz, & Werzberger, 2014). Further, the cost of a green building is the most critical factor affecting its development (Zhang, 2014). Perceived likelihood of a first cost premium certainly acts like a significant barrier to sustainable construction and in some projects, this barrier inhibits sustainability from a business perspective, while in other cases it may completely filter out projects from consideration (Pearce, 2008).

https://doi.org/10.1016/B978-0-12-820062-9.00003-7

The most significant barriers to sustainable design and construction were first cost premium of the project and long payback periods from sustainable practices (Ahn, Pearce, Wang, & Wang, 2013). Similarly, Hwang and Tan (2012) and Zhang, Platten, and Shen (2011) also reported the higher cost premiums as the most significant obstacle in green construction. There are certain researches that justify this claim of initial cost barriers of green buildings. As an example, in a research to identify the costs of green office buildings in Israel, it is reported that the optimum alternative involved an additional cost, ranging between 4% and 12%, whereas under the economical alternative, the additional cost was only 0.12–1.33% (Gabay et al., 2014). Considering residential buildings in China, the analysis results showed that the incorporation of green systems causes the construction costs to increase by 10.77% more than the traditional building, whereas the number of working days only increases by 2 days (Kim, Greene, & Kim, 2014). For the same building compared to a green building with the use of general design, construction techniques, and tools, the use of green building design, construction techniques, and methods will incur additional costs of about 2% of the total investment on average, raising initial costs to 5–10% higher than the ordinary building. Further, green building cost premiums were expected to change according to the type of green certification, the desired level of green building rating, and the nature of the building, and would likely increase with higher levels of certification (Tatari & Kucukvar, 2011). The cost premiums calculated by various research studies varies significantly depending on country, type of building, green certification, and so on.

Davis and Langdon (2007) reported that there is a slight increase in cost whereas the initial impact on construction costs is likely to be in the order of 3–5% for a 5 Star solution in Green Star, with an impact of a further 5% higher for a 6 Star non-iconic design solution. According to the Green Building Council of Australia (2016), there is an average increase in cost of approximately 3% for a Green Star certified project. With these two data, it is possible to depict that approximately after a decade, there is a slight decrease in the cost premium for developing Green Star certified green buildings in Australia.

According to Liu, Guo, and Hu (2014) the incremental costs of the energy efficiency technology applications account for a large proportion of total incremental costs of green buildings, but in return, energy efficiency technology applications on green buildings can bring incremental economic benefits, as well as environmental benefits. In addition, with prices of oil and natural gas skyrocketing in recent years, having energy savings in green buildings every year increases the building value, as occupants are able to recoup their investment in the building within a shorter period of time (Hwang & Tan, 2012). According to Bond (2011), Australia and New Zealand are at the forefront in overlooking these savings for green buildings.

However, according to McAuley (2008), there is a broader economic picture of the indirect economic benefits of green buildings, such as higher public profile, increased productivity, and improved health and morale of employees. According to Qualk and McCown (2009), the concept of green buildings costing more is a part of misconception of lack of project experience. Green buildings can result in significant economic savings by improving employee productivity, increasing benefits from improvements in health and safety, and providing savings from energy, maintenance, and operational costs (Ries, Bilec, Gokhan, & Needy, 2006). Further, according to Ries et al. (2006), based on a research on a green factory building, productivity increased by about 25%, and energy usage decreased by about 30% on a square foot basis compared to a normal conventional factory building. According to McGraw Hill Construction (2013), in new green buildings, operating costs decrease by over 8% over a period of 1 year, and for green retrofits there is a decrease of 9%. All these studies provide significant data to outweigh the initial cost premiums of green buildings when considering the life-cycle savings.

For instance, some green buildings were reported to consume 26% less energy and have demonstrated 13% lower maintenance costs compared to average commercial buildings (Fowler & Rauch, 2006). However, it must be noted that these benefits come with the cost premium that is spent on green buildings compared to conventional buildings. After considering 17 empirical studies, Dwaikat and Ali (2016) illustrated that over 90% of the results reported in the investigated empirical studies fall within the range from − 0.4% to 21%, and very little evidence supports that the green building costs less than a conventional counterpart. However, in this research, it was further pointed out that the literature that addresses the issue of green cost premium is limited and did not yet reach maturity. Additionally, only six publications were classified as academic research, while over 70% of the empirical studies found in the literature are trade publications, in some instances commissioned by governmental bodies (Dwaikat & Ali, 2016).

However, this common assumption of "initial cost premium" is not backed up by recent research and should be questioned, whereas construction professionals need to be informed of the whole life-cycle cost considering the perceived benefits and environmental impact of buildings (Bartlett & Howard, 2000). The perceived economic benefits also must be considered in evaluating the cost of green buildings. Therefore, rather than strictly following the initial cost and investment, it is necessary to look at the broader picture and account for the whole life of the green building, including the benefits in operating cost as well. For this purpose, it is necessary to calculate the life-cycle cost of the green building.

Life-cycle cost can provide motivation for environmental progressive building despite the sometimes higher initial cost (Sterner, 2000). Therefore, life-cycle approach is considered as a valuable approach, enabling operational cost benefits to be evaluated against any initial cost increases (Cole & Sterner, 2000). According to Zuo et al. (2017), tools such as life-cycle costing have been developed to justify extra upfront resources required for green building developments.

3.3 Life-cycle cost for green buildings

The concept of life-cycle costing was initially developed in the mid-1960s to assist the USA Department of Defense in the procurement of military equipment (Epstein, 1996). Further, life-cycle costing was popular among American government agencies in decision-making among different options (Goh & Sun, 2015). In those initial stages onwards, the life-cycle costing approach was formally established and applied, and there have been continued calls to use a discounted present value approach for making economic evaluations of all relevant costs associated with a project investment over its entire life (Goh & Sun, 2015). A general understanding of the essence of the method was that it enables, by calculating the discounted present value of alternative designs of buildings, a comparison of values that transcends problems in comparing projects of differing lives or differing balances between the initial cost of facility procurement and the continuing costs of supporting the facility for its effective operation (Goh & Sun, 2015).

Initially, it is necessary to identify the real meaning of life-cycle costing for buildings. There are many definitions put forward by many researchers in this regard. Basically, it can be identified as a tool for assessing the total cost performance of an asset over time, including the acquisition, operating, maintenance, and disposal costs (Goussous & Al-Refaie, 2014). According to Addis and Talbot (2001 p. 1), life-cycle costing can be identified as.

> ...the present value of the total cost of that asset over its operational life. This includes initial capital cost, finance costs, operational costs, maintenance costs, and the eventual disposal costs of the asset at the end of its life. All future costs and benefits are reduced to present-day values by the use of discounting techniques.

This is very much acceptable in green buildings as well. In general, it is the present value of all the costs associated with the green building over the life-cycle in terms of the present value. Further, life-cycle costing of green buildings can be illustrated as the sum of the incurred costs during its economic life from building pre-decision, design, bidding, construction, completion, and acceptance, until users stop using it. It also includes the sum of research development fees, manufacturing fees, installation fees, operation maintenance fees, and scrap back charges in determining the life-cycle of the project at a predetermined period of validity (Zhang, 2014).

In life-cycle costing, it is very important to identify the necessary costs included in the life-cycle costing calculation. As far as a construction project is considered, there are many types of costs and externalities attached to it. Life-cycle costing involves the systematic consideration of all "relevant" costs and revenues associated with the acquisition and ownership of an asset, and it should not be mixed with other terminologies such as "total cost" and "full cost" (Cole & Sterner, 2000). Usually, in life-cycle costing, the direct and indirect financial costs together with recognized contingent costs are considered in monetary terms, and less quantifiable social costs and external social costs borne by the society are excluded from the study for life-cycle costing (Cole & Sterner, 2000). Similarly, the International Organisation for Standardization [ISO] (2017) also illustrated that life-cycle costs include the construction costs, operational costs, maintenance costs, and end-of-life costs, whereas externalities and social benefits are not included.

Capital cost is the initial investment made for the project. Usually, there are three sub categories to capital cost, namely; purchase cost, acquisition/finance costs and installation/commissioning/training costs (Woodward, 1997). However, in terms of building construction, the capital cost represents the cost of construction as well Operating costs are the expenses incurred within the operation of the buildings. The "residual value" of a structure depends upon whether it is demolished, where the material can be recycled, or more carefully deconstructed to allow structural components to be reused (Gardner, Cruise, Sok, Krishnan, & Santos, 2007). Considering all these types of costs, Table 3.1 summarizes different types of costs at given stages of building life-cycle.

Each of these given costs must be identified in order to calculate the life-cycle cost. Certain costs such as operating and maintenance costs are not available at the time of life-cycle costs calculations and therefore it is necessary predict based on solid data. This involves higher degree of uncertainty. There are many essential parts of the calculation that have to be

TABLE 3.1 Various potential costs in green building life-cycle.

Phase of building life-cycle phase		
Initial stage	**Operational stage**	**Demolition stage**
• Construction costs including material, labor and plant costs • Design fees • Land acquisition costs • Insurance fees (if any)	• Inspection and maintenance costs • Operational costs including utility costs for energy and water • Repair and replacement costs	• Demolition costs • Costs for recovering re-usable and recyclable units • Salvage values (if any)

determined, often on the basis of only scant evidence, data and information and some of this information is of such crucial nature that high quality professional judgment and forecasting is necessary (Ashworth, 1989).

Future costs are usually subject to a level of uncertainty that arises from a variety of factors, including the prediction of the pattern of use of the asset over time; the nature and scale of operating costs; the need for and cost of maintenance activities; the impact of inflation on individual and aggregate costs; the prediction of the length of the asset's useful life; and the significance of future expenditure compared with present day expenditure (Australian National Audit Office, 2001). To forecast these costs, there are cost-estimating techniques that can be used depending on the availability of cost data and the phase in which the life-cycle cost is carried out.

According to Fabrycky and Blanchard (1991), there are three main cost-estimating techniques: estimating by engineering procedures, estimating by analogy, and parametric estimating methods. In estimating by engineering procedures, costs are assigned to each element at the lowest level of design detail, then combined into a total for the product or system (Korpi & Ala-Risku, 2008). This requires detailed data but results in an accurate estimate. In estimating by analogy, as its name already states, the cost estimator draws analogies between different products or their features based on system level or task level (Fabrycky & Blanchard, 1991). This method could be the most inaccurate, depending on the experience and, to a greater extent, expertise of the estimator. Parametric estimation utilizes different statistical techniques and seeks the factors on which the life-cycle costs depend (Korpi & Ala-Risku, 2008). Considering these methods, this research develops the costs from scratch focusing on the first principle. Further details on life-cycle cost calculation are given in a later section (refer Section 5.2).

These costs are incurred in different periods of the life-cycle. Since the timing of costs are different, it is necessary to reflect this in the life-cycle cost calculation (Gluch & Baumann, 2004). For this purpose, the most commonly used technique is the use of discounting and the time value of money, expressed as a discount rate, which depends on inflation, cost of capital, investment opportunities, and personal consumption preferences (Gluch & Baumann, 2004). The discount rate controls the present value of costs over the life-cycle, and the variation of the discount rate changes the impact of costs associated with maintenance, operating, and end-of-life costs, which span over the building life-cycle (Gardner et al., 2007). The important point to note is that discounting techniques are used in order to take account of the time value of money because the present value of a sum of money today depends upon the time at which that money is expended or received (Norman, 1990). The conversion based on the discounting rate is known as net present value (NPV) calculation which is further illustrated in a later chapter (refer Section 5.2).

According to the stage, life-cycle costing of green buildings can be divided into five parts: decision costs, design costs, commissioning costs of construction, operating and maintenance costs, and recycling scrap costs (Yin & Bai, 2014). To achieve the goal of the lowest life-cycle cost of green buildings, the analysis of green building costs should take all factors into consideration and connect hidden environmental and social consumption costs with business costs, which are difficult to quantify (Yin & Bai, 2014).

Apart from the types of costs and the timing of costs at each stage, it is necessary to note that different stakeholders have different "meanings of costs." A significant cost to one stakeholder can be a saving or an income for another stakeholder. As an example, for a project to get certified in Green Building Council, it must be registered and a certain payment must be paid for the services provided by the council (Green Building Council of Australia, 2018). This is one of the initial costs attributed to the developer, yet it is an income for the council. Therefore, when calculating the life-cycle costs, initially it is necessary to define the scope based on the stakeholder that the calculation represents. Based on that classification, it is necessary to identify costs and savings.

3.4 Life-cycle cost models developed for green buildings

There are many studies carried out based on the life-cycle evaluation of different materials and other green initiatives. These research studies focused on many aspects of life-cycle cost within the construction industry. In the literature, there are numerous models focusing on the initial cost of various systems, materials, and so on. However, this research signifies the use of life-cycle cost and, therefore, cost models considering the life-cycle perspective in buildings are considered in this section.

Initially, considering the New Zealand context, Mithraratne and Vale (2004) developed a method to analyze the life-cycle of individual houses in New Zealand based on the embodied and operating energy requirements and life-cycle cost over the useful life of the building. However, in this book, the life-cycle cost is calculated considering the energy rates of the country, and the discount rate is considered to be 5%. Further, this method provided different types of elements and structures for the selection. However, energy is only one criterion of green buildings. Therefore, there is still a need to embed other criteria into this method.

According to Wang, Zmeureanu, and Rivard (2005), selecting better design alternatives satisfying several conflicting criteria—especially, economical and environmental performance—is a challenging task. Therefore, Wang et al. (2005) presented a multi-objective optimization model to assist designers in selecting optimum green building design using life-cycle analysis methodology to evaluate design alternatives for both economical and environmental criteria. However, in this model, the environmental impacts are evaluated in terms of expanded cumulative exergy consumption, which is the sum of exergy consumption due to resource inputs and abatement exergy required to recover the negative impacts due to waste emissions. Similar to the previous method, this model also considered one criterion, namely exergy, which is the available energy for a system. Based on the literature, green buildings have to consider all the other criteria as well. However, this optimization model does not cater to that requirement.

Similarly in 2006, Wang, Rivard, and Zmeureanu (2006) developed methodology to optimize building shapes in a plan using the generic algorithm. In this model, life-cycle cost and life-cycle environmental impact are the two objective functions used to evaluate the performance of a green building design. Similar to the previous model, this model also focuses on environmental performance excluding social criteria, such as IEQ.

Gu et al. (2007) proposed a method named life-cycle green cost assessment (LCGCA) to evaluate building environmental load and economic performance throughout its life-cycle comprehensively. However, this method only focused on the environmental performance of the building as opposed to economic performance. Similarly, Verbeeck and Hens (2007) also developed an optimization model for extremely low-energy dwellings, taking into account energy use, environmental impact, and financial costs over the life-cycle of the buildings. According to Section 2.4, green buildings focused on the triple bottom-line construct, and this model excludes the social parameter from this method. Therefore, green initiatives such as IEQ are not evaluated in terms of costs within the building life-cycle.

Hasan, Vuolle, and Sirén (2008) developed an optimization program for detached houses to minimize the life-cycle cost. However, this optimisation program optimized values of five selected design variables in the building construction and HVAC system. These variables are insulation thickness of the external wall, roof, and floor, and two discrete variables: the U-value of the windows and type of heat recovery. Therefore, this program minimizes the life-cycle cost focusing on the building elements and the HVAC system. Once again, other criteria such as water efficiency and energy are not considered in this model even though the life-cycle cost is minimized.

Similarly, Chai, Hu, Peng, and Wang (2010) developed an incremental economy model for green buildings for water savings. This model is developed by analyzing the composition of direct incremental cost within the building life-cycle. Therefore, the model used both initial costs and the future costs that will occur in the life-cycle of the building, and the future costs are discounted back using the NPV technique. Further, the model is validated considering a water-saving demonstration project for green building in western China (Chai et al., 2010). This model is developed to calculate the life-cycle costs for water savings. However, according to the literature (refer to Section 2.4), there are many other criteria that need to be achieved in green buildings. The rest of the criteria are not considered in this model, which is one of the major limitations in this study.

Tuhus-Dubrow and Krarti (2010) developed a simulation-optimization tool to optimize building shape and building envelope features considering the lowest life-cycle cost on energy. This model selects the optimum building envelope for the lowest life-cycle cost. This tool couples a generic algorithm to a building energy simulation engine to select optimal values to minimize energy use for residential buildings. This model also considers energy optimization through the changes in building envelopes. Similarly, Fesanghary, Asadi, and Geem (2012) also developed a multi-objective optimization model to minimize the life-cycle cost and carbon dioxide equivalent (CO_2-eq) emissions of the buildings. Further, this model included various building envelope parameters and design variables. However, the aim of the model is to optimize the carbon emission with lowest life-cycle cost.

Bichiou and Krarti (2011) developed an optimization model to select both building envelope features and heating and air conditioning system design and operation settings to minimize the life-cycle cost. In this model, the building design features are determined to minimize the life-cycle costs. It is found that the optimal selection can reduce life-cycle costs by 10–25%, depending on the climate and type of home (Bichiou & Krarti, 2011).

There are certain life-cycle cost models focusing on specific criteria of green buildings. He, Meng, Gao, Li, and Li (2013) developed an evaluation software to select the optimal design scheme using entropy decision-making method to support the decision makers in comparing the different design schemes and to calculate the best material to the selected scheme considering the life-cycle cost. Once again, in this software the only focus is given to the material selection based on the life-cycle cost. All the remaining criteria, such as water efficiency, energy, and IEQ, are not supported through this software.

Dwaikat and Ali (2014) presented a conceptual model considering the life-cycle cost performance of the building and earned value management (EVM) approach. However, this EVM approach can be applied throughout the green building's operating phase and also investigate the cost patterns and characteristics in both the building construction and operating phase. This conceptual model is applied to life-cycle cost control rather than in the decision-making process.

Robati, McCarthy, and Kokogiannakis (2018) conducted a research to reflect the environmental impact and the building cost into the decision-making process in sustainable structural design. This study proposed a method that integrates and considers the environmental cost and building cost in the structural design process (Robati et al., 2018). In this research, Robati et al. (2018) focused on the building material cost, construction methods, and amount of embodied carbon emissions during the life-cycle of buildings for two types of slabs and two structural materials. Based on the research findings, Robati et al. (2018) found that an appropriate selection of construction forms and type of concrete can save up to 7% of the cost of material consumption, 5% of the total energy consumption expense, and 5% of the CO_2-eq emissions of the building across all five major cities in Australia. However, this model only focused on building element and only materials are considered in terms of cost calculation.

When considering all these life-cycle cost models, it is necessary to identify that all these models only focused on one criterion of the building. According to the literature, green buildings are environmentally, socially, and economically sustainable buildings. However, these three aspects are not holistically included in any of these life-cycle cost models. Further, according to Section 2.4.1, green buildings are evaluated based on the green building rating tools. Therefore, these green building rating tools are considered as yardsticks to evaluate the green initiatives in a building. Each of these rating tools has a similar set of key criteria. However, all these life-cycle cost models focus on either one or two of these key criteria, completely ignoring the rest of the green initiatives. Therefore, there is a clear lack of research in developing a life-cycle cost model to cater to all the key criteria, focusing on the green building as a whole system.

3.5 Summary

Cost premium is widely discussed in the green building industry. Many research studies identify that there are considerable percentage increases in the initial cost of green buildings, and certain studies identify slight increments compared to conventional buildings. On the other hand, when considering the operating costs, such as water and electricity, there are significant savings in green buildings over the life-cycle of the buildings. Hence, when evaluating the "cost" of the buildings, it is necessary to consider the building life-cycle costs rather than the initial cost of the buildings to obtain a broader understanding. As a result, life-cycle costing of green buildings is considered a significant study.

In simple terms, life-cycle costing refers to the costs incurred over the life-cycle of a building. When a building is considered, there are a lot of costs associated with that. In life-cycle costing, it is necessary to initially demarcate which costs are included in the study. Generally, in life-cycle costing, the main costs can be categorized as initial cost, operating costs, maintenance costs, and demolition of the building at the end of the life-cycle. When all these costs are calculated with the use of estimating techniques, life-cycle cost can be obtained. Since these costs are spread across the lifespan of the building life-cycle, when calculating the cost, the time value of money must be considered. This is embedded in the life-cycle cost calculation by using discounting techniques. Generally, the NPV is calculated to get the present value of the future costs. After all these costs are calculated, finally, life-cycle costing is established.

So far, there are many life-cycle cost models developed to identify optimum solutions. However, all these life-cycle cost model focus on one or two criteria of green building whereas the majority of the green building criteria are not supported. Therefore, the literature clearly identified a lack of research in developing a life-cycle cost model considering all the key criteria required in a green building.

References

Addis, B., & Talbot, R. (2001). Sustainable construction procurement—A guide to delivering environmentally responsible projects. UK: Construction Industry Research and Information Association (CIRIA).

Ahn, Y. H., Pearce, A. R., Wang, Y., & Wang, G. (2013). Drivers and barriers of sustainable design and construction: The perception of green building experience. *International Journal of Sustainable Building Technology and Urban Development*, *4*(1), 35–45. https://doi.org/10.1080/2093761X.2012.759887.

Ashworth, A. (1989). Life-cycle costing: A practice tool? *Cost Engineering*, *31*(3), 8.

Australian National Audit Office. (2001). Life cycle costing—Better practice guide. Retrieved from http://www.anao.gov.au/uploads/documents/Life_Cycle_Costing.pdf.

Bartlett, E., & Howard, N. (2000). Informing the decision makers on the cost and value of green building. *Building Research & Information*, *28*(5–6), 315–324. https://doi.org/10.1080/096132100418474.

Bichiou, Y., & Krarti, M. (2011). Optimization of envelope and HVAC systems selection for residential buildings. *Energy and Buildings*, *43*(12), 3373–3382. https://doi.org/10.1016/j.enbuild.2011.08.031.

Bond, S. (2011). Barriers and drivers to green buildings in Australia and New Zealand. *Journal of Property Investment & Finance*, *29*(4/5), 494–509. https://doi.org/10.1108/14635781111150367.

Chai, H. X., Hu, X. B., Peng, S. J., & Wang, T. Y. (2010). Incremental-cost economy model in full life cycle of water-saving projects for green building. *Huanan Ligong Daxue Xuebao/Journal of South China University of Technology (Natural Science)*, *38*(11), 59–63. https://doi.org/10.3969/j.issn.1000-565X.2010.11.011.

Cole, R., & Sterner, E. (2000). Reconciling theory and practice of life-cycle costing. *Building Research & Information*, *28*(5–6), 368–375. https://doi.org/10.1080/096132100418519.

Davis, & Langdon. (2007). The cost and benefits of achieving green buildings. Retrieved from http://www.usgbc.org/Docs/Archive/General/Docs2583.pdf.

Dwaikat, L. N., & Ali, K. N. (2014). Green buildings' actual life cycle cost control: A framework for investigation. In *Paper presented at the 13th management in construction research association conference and annual general meeting. International Islamic University of Malaysia.*

Dwaikat, L. N., & Ali, K. N. (2016). Green buildings cost premium: A review of empirical evidence. *Energy and Buildings*, *110*(Suppl. C), 396–403. https://doi.org/10.1016/j.enbuild.2015.11.021.

Epstein, M. J. (1996). Measuring corporate environmental performance: Best practices for costing and managing an effective environmental strategy. Chicago, IL: Irwin Professional Publishing.

Fabrycky, W. J., & Blanchard, B. S. (1991). Life cycle cost and economic analysis. Englewood Cliffs, NJ: Prentice-Hall.

Fesanghary, M., Asadi, S., & Geem, Z. W. (2012). Design of low-emission and energy-efficient residential buildings using a multi-objective optimization algorithm. *Building and Environment*, *49*(1), 245–250. https://doi.org/10.1016/j.buildenv.2011.09.030.

Fowler, K. M., & Rauch, E. M. (2006). Sustainable building rating systems summary *(PNNL-15858)*. Retrieved from http://www.pnl.gov/main/publications/external/technical_reports/PNNL-15858.pdf.

Gabay, H., Meir, I. A., Schwartz, M., & Werzberger, E. (2014). Cost-benefit analysis of green buildings: An Israeli office buildings case study. *Energy and Buildings*, *76*, 558–564. https://doi.org/10.1016/j.enbuild.2014.02.027.

Gardner, L., Cruise, R. B., Sok, C. P., Krishnan, K., & Santos, J. M. D. (2007). Life-cycle costing of metallic structures. *Proceedings of the Institution of Civil Engineers—Engineering Sustainability*, *160*(4), 167–177. https://doi.org/10.1680/ensu.2007.160.4.167.

Gluch, P., & Baumann, H. (2004). The life cycle costing (LCC) approach: A conceptual discussion of its usefulness for environmental decision-making. *Building and Environment*, *39*(5), 571–580. https://doi.org/10.1016/j.buildenv.2003.10.008.

Goh, B. H., & Sun, Y. (2015). The development of life-cycle costing for buildings. *Building Research & Information*, 1–15. https://doi.org/10.1080/09613218.2014.993566.

Goussous, J., & Al-Refaie, A. (2014). Evaluation of a green building design using LCC and AHP techniques. *Life Science Journal*, *11*(8s).

Green Building Council of Australia. (2016). Green star financial transparency. Retrieved from Australia: https://gbca-web.s3.amazonaws.com/media/documents/gbca-research-paper---financial-transparency-2016.pdf.

Green Building Council of Australia. (2018). About GBCA. Retrieved from http://www.gbca.org.au/about/.

Gu, L., Gu, D., Lin, B., Huang, M., Gai, J., & Zhu, Y. (2007). Life cycle green cost assessment method for green building design. Paper presented at the building simulations 2007.

Hasan, A., Vuolle, M., & Sirén, K. (2008). Minimisation of life cycle cost of a detached house using combined simulation and optimisation. *Building and Environment*, *43*(12), 2022–2034. https://doi.org/10.1016/j.buildenv.2007.12.003.

He, W. Y., Meng, F. R., Gao, X. Z., Li, Z. W., & Li, K. (2013). Research on life cycle cost modular evaluation system of green building materials: Based on entropy decision making method. *Advanced Materials Research*, *798*, 1152–1157. https://doi.org/10.4028/www.scientific.net/AMR.798-799.1152.

Hwang, B. G., & Tan, J. S. (2012). Green building project management: Obstacles and solutions for sustainable development. *Sustainable Development*, *20*(5), 335–349.

International Organisation for Standardization [ISO]. (2017). Buildings and constructed assets—Service life planning Part 5: Life cycle costing (Vol. ISO 15686-5). Switzerland: International Standard Organisation [ISO].

Kim, J.-L., Greene, M., & Kim, S. (2014). Cost comparative analysis of a new green building code for residential project development. *Journal of Construction Engineering and Management*, *140*(5). https://doi.org/10.1061/(ASCE)CO.1943-7862.0000833. 05014002-05014001–05014002-05014010.

Korpi, E., & Ala-Risku, T. (2008). Life cycle costing: A review of published case studies. *Managerial Auditing Journal*, *23*(3), 240–261. https://doi.org/10.1108/02686900810857703.

Liu, Y., Guo, X., & Hu, F. (2014). Cost-benefit analysis on green building energy efficiency technology application: A case in China. *Energy and Buildings*, *82*, 37–46. https://doi.org/10.1016/j.enbuild.2014.07.008.

McAuley, T. (2008). Understanding green building cost and value. *AACE International Transactions*. HP41.

McGraw Hill Construction. (2013). World green building trends. Retrieved from http://www.worldgbc.org/files/8613/6295/6420/World_Green_Building_Trends_SmartMarket_Report_2013.pdf.

Mithraratne, N., & Vale, B. (2004). Life cycle analysis model for New Zealand houses. *Building and Environment*, *39*(4), 483–492. https://doi.org/10.1016/j.buildenv.2003.09.008.

Norman, G. (1990). Life cycle costing. *Property Management*, *8*(4), 344–356. https://doi.org/10.1108/EUM0000000003380.

Pearce, A. R. (2008). Sustainable capital projects: Leapfrogging the first cost barrier. *Civil Engineering and Environmental Systems*, *25*(4), 291–300. https://doi.org/10.1080/10286600802002973.

Qualk, J. D., & McCown, P. (2009). The cost-effectiveness of building 'green'. *Heating/Piping/Air Conditioning Engineering: HPAC*, *81*(10), 18–20. 22–23.

Ries, R., Bilec, M. M., Gokhan, N. M., & Needy, K. L. (2006). The economic benefits of green buildings: A comprehensive case study. *The Engineering Economist*, *51*(3), 259–295. https://doi.org/10.1080/00137910600865469.

Robati, M., McCarthy, T. J., & Kokogiannakis, G. (2018). Integrated life cycle cost method for sustainable structural design by focusing on a benchmark office building in Australia. *Energy and Buildings*, *166*, 525–537. https://doi.org/10.1016/j.enbuild.2018.02.034.

Sterner, E. (2000). Life-cycle costing and its use in the Swedish building sector. *Building Research & Information*, *28*(5–6), 387–393. https://doi.org/10.1080/096132100418537.

Tatari, O., & Kucukvar, M. (2011). Cost premium prediction of certified green buildings: A neural network approach. *Building and Environment*, *46*(5), 1081–1086. https://doi.org/10.1016/j.buildenv.2010.11.009.

Tuhus-Dubrow, D., & Krarti, M. (2010). Genetic-algorithm based approach to optimize building envelope design for residential buildings. *Building and Environment*, *45*(7), 1574–1581. https://doi.org/10.1016/j.buildenv.2010.01.005.

Verbeeck, G., & Hens, H. (2007). Life cycle optimization of extremely low energy dwellings. *Journal of Building Physics*, *31*(2), 143–177. https://doi.org/10.1177/1744259107079880.

Wang, W., Rivard, H., & Zmeureanu, R. (2006). Floor shape optimization for green building design. *Advanced Engineering Informatics*, *20*(4), 363–378. https://doi.org/10.1016/j.aei.2006.07.001.

Wang, W., Zmeureanu, R., & Rivard, H. (2005). Applying multi-objective genetic algorithms in green building design optimization. *Building and Environment*, *40*(11), 1512–1525. https://doi.org/10.1016/j.buildenv.2004.11.017.

Woodward, D. G. (1997). Life cycle costing—Theory, information acquisition and application. *International Journal of Project Management*, *15*(6), 335–344. https://doi.org/10.1016/S0263-7863(96)00089-0.

Yin, Y. L., & Bai, J. (2014). Study on cost of green building based on the life cycle theory. *Applied Mechanics and Materials*, *587-589*, 228–231. https://doi.org/10.4028/www.scientific.net/AMM.587-589.228.

Zhang, J. W. (2014). Cost, efficiency and hygiene—Three reflections on green building. *Applied Mechanics and Materials*, *587-589*, 203–207. https://doi.org/10.4028/www.scientific.net/AMM.587-589.203.

Zhang, X., Platten, A., & Shen, L. (2011). Green property development practice in China: Costs and barriers. *Building and Environment*, *46*(11), 2153–2160. https://doi.org/10.1016/j.buildenv.2011.04.031.

Zuo, J., Pullen, S., Rameezdeen, R., Bennetts, H., Wang, Y., Mao, G., … Duan, H. (2017). Green building evaluation from a life-cycle perspective in Australia: A critical review. *Renewable and Sustainable Energy Reviews*, *70*, 358–368. https://doi.org/10.1016/j.rser.2016.11.251.

CHAPTER

4

Green Star Australia credit classification

4.1 Introduction

This book is based on the Green Star Design and As-Built version 1.1 (Green Building Council of Australia, 2015) rating tool. This rating tool is for new construction and major refurbishments. However, this book focused on new constructions and, specifically, commercial office buildings. The Green Star Design and As-Built version 1.1 has many credits. The total number of credits amount to 100. This rating tool provides different types of credits. Therefore, before calculating the life-cycle cost for credits, it was necessary to have a clear understanding of the credits. Therefore, the authors carried out a thorough analysis of each of the credits. The analysis revealed various types of credits such as the ones given below:

- Conditional credits that are mandatory to achieve when claiming credit points under certain sub-criteria
- Additional credits with additional credit points that are claimed when a building exceeds the required level of performance
- Same credits that can be achieved using different pathways
- Different options that are available for achieving same credits, depending on the type of system used inside a building
- Different credits that focus on the type of building
- Credits that can be achieved by obtaining the certification of another assessment tool
- Credits that offer various choices.

Based on the analysis of the different types of credits, certain credits can be eliminated from the research, and certain credits can be selected. This chapter of this book presents an analysis of various types of credits. Finally, this chapter illustrates the eliminated credits with reasoning and presents the credits chosen to further proceed with the life-cycle cost calculation.

https://doi.org/10.1016/B978-0-12-820062-9.00004-9

Once the credit selections are carried out, the upcoming chapters in this book would further analyze the selected credits and establish dependencies among these credits in terms of cost and credits selections. Section 4.4 of this chapter illustrates the dependencies among credits.

4.2 Methodology for credit classification

This chapter focuses on analyzing the different credits and credits points. After the analysis, certain credits are eliminated. Upon elimination of credits, the remaining credits are further analyzed to develop the dependencies. Therefore, there are two sections for the methodology, as illustrated in Fig. 4.1 below.

Upon analyzing the credits, authors identified eight credit types as given in Fig. 4.1. Afterwards, certain credits are eliminated from the study due to various solid reasons which are illustrated in Section 4.3.8 of this chapter.

4.3 Green star criteria, credits, and credit points

Prior to the credit classification, it is necessary to understand the terminology used in credit classification. The Green Star Design and As-Built version 1.1 rating tool assesses the sustainability outcomes from design, new construction, and major refurbishments of buildings (Green Building Council of Australia, 2015). There are nine key criteria for Green Star ratings: management (M), indoor environment quality (IEQ), energy (E), transport (T), water (W), material (Mat), land use and ecology (L), and emissions (Em) and innovation (I). The key criterion of innovation provides a platform to reward exceptional performance in green buildings. It does not provide any specific guideline to achieve the credit points; however, it allows new innovative technologies. Therefore, the key criterion of innovation is not considered in this book. The model only considers eight key criteria excluding innovation. Each key criterion includes detailed sub-criteria and credits, and each of these credits can score specific credit points that are summed to achieve the desired green building rating. Fig. 4.2 illustrates each of these categories.

The "key criteria" refers to the nine key categories under which all the other categories are listed. In each key criterion, several "sub-criteria" are identified by the rating tools. Subsequently, each of these sub criteria are further divided into "credits." Sub-criteria identify the attributes required for each key criterion and credits illustrate the requirements for achieving green building status. Finally, "credits points" provide the points allocated for each credit. According to Fig. 4.2, the IEQ and material (Mat) refer to key criteria and "Indoor Air Quality" and "Responsible Building Material" refer to sub-criteria, respectively. Additionally, three credits are listed for each of these sub-criteria; credit points corresponding to each of these credits are also listed in Fig. 4.2.

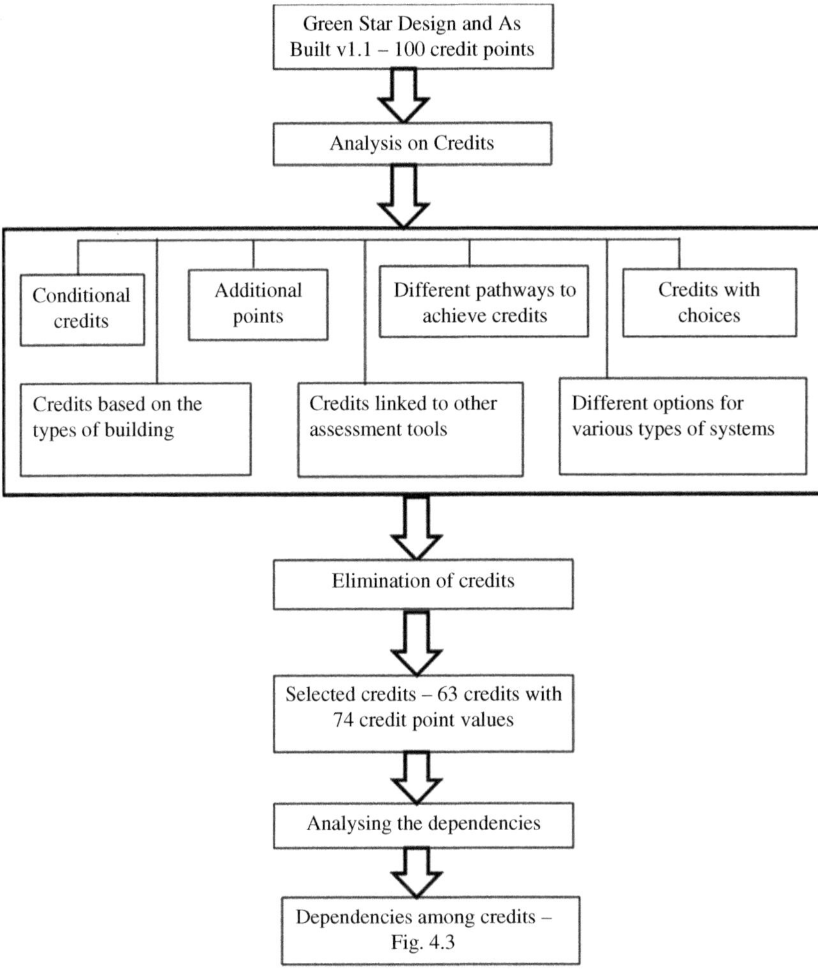

FIG. 4.1 Research method for this chapter.

There are different types of Green Star credits and credit points, such as conditional credits, credits with additional credit points, and credits with different pathways. Therefore, at the outset, all these credits and credit points are analyzed.

4.3.1 Conditional credits

As shown in Fig. 4.2, there are several credits listed under each sub-criterion. However, it is essential to meet certain conditional requirements under certain sub-criteria. These conditional requirements are represented by credits that do not have any credit points. For example, the "M6.0 Metering" credit is a conditional requirement, and, therefore, the "M6.0

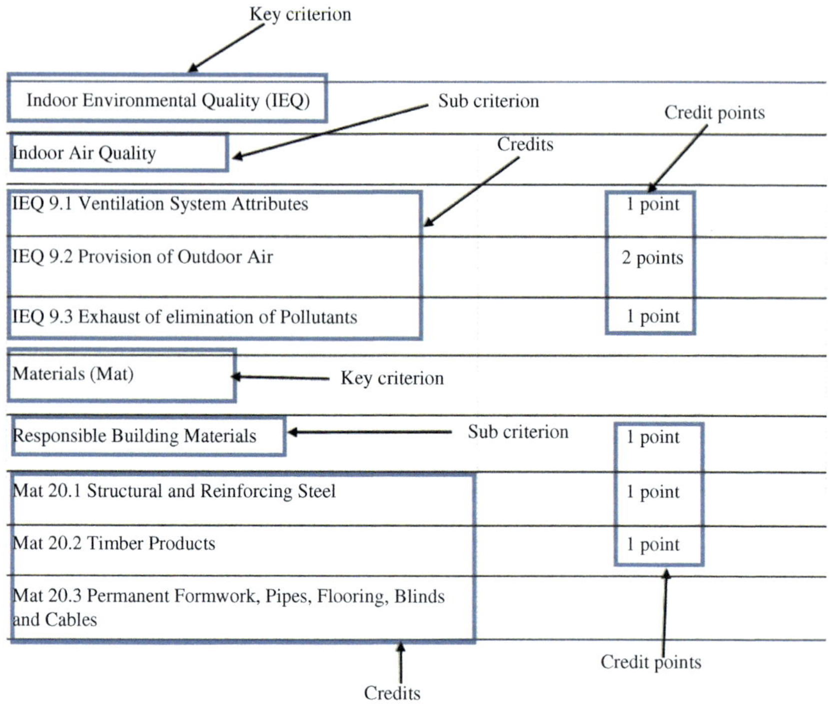

FIG. 4.2 Classification of key criteria, sub-criteria, credits, and credit points.

Metering" must be met to obtain the "M6.1 Monitoring Systems" credit. Conditional requirements are also present under the key criteria such as management, IEQ, energy, land use and ecology, and emissions. These conditional credits are very important for the development of the model because, if the model selects any of the credits listed in the sub-criteria with a conditional requirement, then it would be essential to select the conditional credits for the calculation of the life-cycle cost. Additionally, these are the only credits that are without credit points. Table 4.1 reports the conditional credits in Green Star Design and As-Built version 1.1.

According to Table 4.1, if the user achieves any of the credits that fall under the sub-criteria, such as commissioning and tuning, metering and monitoring, construction environmental management, lighting comfort, visual comfort, ecological value, sustainable sites, and light pollution, then it will be necessary to fulfill the conditional credits, respectively.

4.3.2 Additional points

Certain additional points are offered if a building performs beyond a certain level. These points do not fall under the key criterion

TABLE 4.1 Conditional credits.

Key criteria	Sub-criteria	Credit	Credit point
Management (M)	Commissioning and Tuning	M2.0 Environmental Performance Targets	Required
	Metering and Monitoring	M6.0 Metering	Required
	Construction Environmental Management	M7.0 Environmental Management Plan	Required
Indoor environment quality (IEQ)	Lighting Comfort	IEQ11.0 Minimum Lighting Comfort	Required
	Visual Comfort	IEQ12.0 Glare Reduction	Required
Land use and ecology (L)	Ecological Value	L23.0 Endangered, Threatened, or Vulnerable Species	Required
	Sustainable Sites	L24.0 Conditional Requirement	Required
Emissions (Em)	Light Pollution	Em27.0 Light Pollution to Neighboring Bodies	Required

of "Innovation." However, these credits are specified in the Green Star Design and As-Built version 1.1. It is essential to achieve certain credits in order to achieve these credit points. For example, prior to achieving the "M2.4 Independent Commissioning Agent," a user must achieve any of the "M2.1 Services and Maintainability Review," "M2.2 Building Commissioning," or "M2.3 Building Systems Tuning" credits. Once any of these credits are fulfilled, the user can consider achieving the additional credit points given. Table 4.2 reports the additional credits points available in the Green Star Design and As-Built version 1.1. According to Table 4.2, the key criteria management, IEQ, and emissions allow additional credits for exceptional building performance. These credits are essential for developing the cost model and must be met to allow the model to select additional credits.

4.3.3 Different pathways to achieve credits

In Green Star Design and As-Built version 1.1, different pathways are available that facilitate the achievement of relevant credits. The most commonly used pathways are the "prescriptive pathway" and the "performance pathway." In "prescriptive pathway," the rating tools illustrate

TABLE 4.2 Additional credit points.

Key criteria	Credit	Credit point	Credit requirements
Management (M)	M2.4 Independent Commissioning Agent	1 point	At least one of the requirements M2.1 Services and Maintainability review M2.2 Building Commissioning M2.3 Building Systems Tuning
Indoor environment quality (IEQ)	IEQ14.2 Advanced Thermal Comfort	1 point	IEQ14.1 Thermal Comfort
Emissions (Em)	Em26.2 Reduced Pollution Targets	1 point	Em26.1 Reduced Peak Discharge

the procedure that needs to be followed. For example, two pathways are available in the "Operational Waste" sub-criteria, namely, the "M8A Performance Pathway—Specialist Plan" and the "M8B Prescriptive Pathway—Facilities." In the performance pathway, a waste professional specialist prepares and implements an operational and waste management plan for the project, thereby influencing the amount of waste recycled and generated by the tenants, occupants, and visitors. In this pathway, the waste professional specialist can put forward numerous solutions for waste management. Therefore, it is impossible to identify all these possible solutions for waste management and calculate the lifecycle cost for each of the possible solutions. However, the prescriptive pathway illustrates the exact requirements that should be carried out to achieve the credits. Therefore, among the prescriptive pathway and performance pathway, this book only selects the prescriptive pathway for calculations. In the "Operational Waste" sub-criteria, both these pathways provide one credit point each. However, in certain credits, these two pathways provide different values for the credit points. For example, in the "Greenhouse Gas Emission" sub-criteria, the prescriptive pathway allocated only five credit points, whereas the "E15E GHG Emissions Reduction—Modelled Performance," which is the performance pathway, allocated 20 credit points.

Table 4.3 reports the credits and sub-criteria with different pathways that are required to achieve credit points. According to Table 4.3, for certain credits, there are pathways, other than prescriptive and performance pathways that can be followed to achieve the required credit points. The available pathways are based on different techniques used to achieve the required green outcome. For example, in the credit "IEQ13.1.2 Carpets," there are two pathways that can be followed to obtain the credit point. In this credit, either the product is certified under a recognized product

TABLE 4.3 Credits/sub criteria with different pathways.

Key criteria	Credit/sub criteria	Credit point	Pathways
Management (M)	M5.2 Environmental Building Performance	1 point	M5.1.1A Building Performance Metrics M5.1.1B Certified Operational Performance Ratings
	Operational Waste	1 point	M8A Performance Pathway M8B Prescriptive Pathway
Indoor environment quality (IEQ)	IEQ11.1.2 Glare reduction	1 point	IEQ11.1.2A Prescriptive Method IEQ11.1.2B Prescriptive Method IEQ11.1.2C Performance Method
	IEQ11.2A Surface Illuminance	1 point	IEQ11.2A Prescriptive Method IEQ11.2B Performance Method IEQ11.2C Residential Spaces
	IEQ12.1 Daylight	2 points	IEQ12.1A Prescriptive Method IEQ12.1B Compliance using Daylight Factor IEQ12.1B Compliance using Daylight Autonomy
	IEQ13.1.2 Carpets	1 point	IEQ13.1.2A Product Certification IEQ13.1.2A Laboratory Testing
Energy (E)	Greenhouse Gas Emissions	Up to 20 points	E15A GHG Emissions Reduction—Prescriptive Pathway E15B GHG Emissions Reduction—NatHERS E15C GHG Emissions Reduction—BASIX E15D GHG Emissions Reduction—NABERS Energy Commitment Agreement E15E GHG Emissions Reduction—Modeled Performance
	Greenhouse Gas Emissions	Up to 2 points	E16A Prescriptive Pathway: On-site Energy Generation E16B Modeled Performance Pathways: Reference Building
Transport (T)	Sustainable Transport	Up to 10 points	T17A Performance Pathway T17B Prescriptive Pathway
Water (W)	Potable Water	Up to 12 points	W18A Performance Pathway W18B Prescriptive Pathway

Continued

TABLE 4.3 Credits/sub criteria with different pathways—cont'd

Key criteria	Credit/sub criteria	Credit point	Pathways
Materials (Mat)	Life Cycle Impacts	Up to 7 points	Mat 19A Performance Pathway—Life Cycle Assessment Mat 19B Prescriptive Pathway—Life Cycle Impacts
	Life Cycle Impacts—Steel	Up to 2 points	Mat 19B.2A Reduced Mass of Steel Framing Mat 19B.2B Reduced Mass of Steel Reinforcement
	Responsible Building Materials	1 point	Mat20.1A Structural and Reinforcing Steel—For steel-framed buildings Mat20.1B Structural and Reinforcing Steel—For concrete-framed buildings
	Construction and Demolition Waste	1 point	Mat22A Reduction of Construction and demolition Waste—Fixed Benchmark Mat22A Reduction of Construction and demolition Waste—Percentage Benchmark

certification scheme according to "IEQ13.1.2A Product Certification" or the product is tested for the given standard according to "IEQ13.1.2A Laboratory Testing." In such instances, the user can determine the pathway that must be used to comply with requirement; it must be noted that the credit points would be the same irrespective of the pathway that is selected.

Apart from these given pathways, Green Star requirements can be achieved by using other green assessment schemes available in Australia. The certifications are part of other assessment tools that can be used to achieve credits in Green Star. According to Table 4.3, the "Greenhouse Gas Emissions" (GHGs) sub-criteria allow the use of Nationwide House Energy Rating Scheme (NatHERS), National Australian Built Environment Rating System (NABERS), and Building Sustainability Index (BASIX) ratings for achieving credit points. This is further discussed in Section 4.3.6.

There are other pathways that can be selected, depending on the requirement of the building. There are two pathways for the "Life Cycle Impacts—Steel" sub-criteria. The "Mat 19B.2A Reduced Mass of Steel Framing" pathway is for steel-framed buildings and the "Mat 19B.2B Reduced Mass of Steel Reinforcement" pathway is for concrete-framed buildings. Therefore, depending on the structure of the building, the user must choose the relevant pathway. Similarly, different pathways can be followed for achieving a credit, depending on the method or strategy used to achieve that credit. For example, the "Construction and Demolition Waste" sub-criteria has two pathways. One pathway focuses on minimizing the total amount of waste sent to a landfill, and the other pathway focuses on diverting a significant amount of waste from going to landfill. Therefore, the users can select the pathway depending on the strategy suitable for the user.

4.3.4 Different options for various types of systems

Building services form an integral part of any building. There are services such as electrical system; HVAC system; and water system. Each of these services has specified systems that are specific to every building. Therefore, based on the specific system, there are separate installations and requirements. For example, if a building has a mechanically-ventilated air conditioning system, then the requirements and the standards that the building must follow would differ from a naturally-ventilated building. Considering this, the Green Star Design and As-Built version 1.1 stipulates different requirements to achieve Green Star points, as given in Table 4.4. According to Table 4.4, there are different requirements for achieving the Green Star points, depending on the type of ventilation provided in the building.

TABLE 4.4 Different options for various types of systems.

Key criteria	Credit	Credit point	Type of system
Indoor environment quality (IEQ)	IEQ9.2 Provision of Outdoor Air	2 points	Separate provisions for mechanically-ventilated spaces and naturally-ventilated spaces
	IEQ14.1 Thermal Comfort	1 point	Separate provisions for mechanically-ventilated spaces, naturally ventilated spaces, and residential spaces
	IEQ14.2 Advanced Thermal Comfort	1 point	Separate provisions for mechanically-ventilated spaces, naturally-ventilated spaces, and residential spaces
Water (W)	W18B.3 Heat Rejection	2 points	Separate provisions for mechanically-ventilated spaces and naturally-ventilated spaces

4.3.5 Credits based on the type of building

The requirements of the building change with the classification of the building. Therefore, the Green Star Design and As-Built version 1.1 has specified certain different standards to suit various buildings. Table 4.5 reports all the credits with varying requirements based on the building type. This book is based on office buildings. Therefore, requirements for office buildings are selected in each of the credits listed below.

4.3.6 Credits linked to another assessment tool

Green Star Australia has a set of rating tools for green buildings (refer Section 2.4.2). Furthermore, there are other assessment tools used in Australia for assessing the performance of buildings in terms of water efficiency and energy efficiency, among others. These assessments also provide certifications that can be used to obtain credits in Green Star Design and As-Built version 1.1. Table 4.6 reports all the credits linked to other assessment tools. However, in this book, these credit points are not considered primarily due to the complexity of analyzing the requirements of other assessment tools. Therefore, in such instances, this book would select other pathways if available.

According to Table 4.6, other assessment and rating tools used in the Green Star Design and As-Built version 1.1 are Green Star Performance, NABERS, NatHERS and BASIX.

TABLE 4.5 Credits based on the type of building.

Key criteria	Credit	Credit point	Remarks
Indoor environment quality (IEQ)	IEQ11.1 General Illuminance and Glare reduction	1 point	There are different standards for different building-types. "Office spaces" is selected for this book
Energy (E)	E17B.2 Reduced Car Parking Provision	1 point	There are different requirements for different building types. "Office" is selected for this book
	E17B.4.1 Active transport Facilities	1 point	There are different requirements for different building types. Class 3 to 9 that represent office buildings are selected for this book

TABLE 4.6 Credits linked to another assessment tool.

Key criteria	Credit	Credit point	Assessment tool
Management (M)	M5.1.1B Certified Operational Performance Ratings	1 point	Green Star Performance rating tool Or NABERS
Energy (E)	E15B GHG Emissions Reduction—NatHERS E15C GHG Emissions Reduction—BASIX E15D GHG Emissions Reduction—NABERS Energy Commitment Agreement	Up to 20 points	NatHERS BASIX NABERS Energy Commitment Agreement

4.3.7 Credits with choices

In the Green Star Design and As-Built version 1.1, there are certain credits with choices, and the user is required to make the most suitable selections. Table 4.7 reports all the credits with choices.

The main difference between this choice and using different pathways, as illustrated in Section 4.3.3, is that, while making choices, the users can choose from a given list of credits to fulfill the requirements for achieving the credit point, whereas the users have to choose either of the available pathways while selecting a pathway. For example, in the "E15A Prescriptive Pathway," 8 credit points are suggested in the rating tool, and the users can choose up to 5 credit points. Therefore, in this type of credits, the user can choose the most preferred solutions from a given list and still achieve the maximum available credit points. According to Table 4.7, the "M5.2 Environmental Building Performance," "E15A Prescriptive Pathway," "Mat19B Prescriptive Pathway—Life Cycle Impacts," and "Mat21 Product Transparency and Sustainability" are the credits with choices.

4.3.8 Credits eliminated

Previous sections (see Sections 4.3.1–4.3.7) discussed different types of credits available in the Green Star Design and As-Built version 1.1. However, certain types of credits are eliminated from this book due to various reasons. Table 4.8, reports all the credits eliminated from this book.

According to Section 4.3.3, there are two types of pathways that can be followed to achieve credits. The performance pathway that must be selected to achieve a particular credit depends on the user, and the user has

TABLE 4.7 Credits with choices.

Key criteria	Credit	Credit point	Choices
Management (M)	M5.2 Environmental Building Performance	1 point	At least two of the following: Greenhouse gas emissions Potable water usage Operational waste Indoor environment quality
Energy (E)	E15A Prescriptive Pathway	5 points	Out of 8 points, 5 points can be achieved: Building envelope Glazing Lighting Ventilation and air conditioning Domestic hot water systems Building sealing
Materials (Mat)	Mat19B Prescriptive Pathway—Life Cycle Impacts	Up to 5 points	Out of 8 points, 5 points can be achieved: Mat19B.1 Life Cycle Impacts: Concrete Mat19B.1 Life Cycle Impacts: Steel Mat19B.1 Life Cycle Impacts: Building Reuse
	Mat21 Product Transparency and Sustainability	Up to 3 points	Any of the following initiatives must be selected: **A.** Reused products **B.** Recycled content products **C.** Environmental Product Declarations **D.** Third-party Certification **E.** Stewardship Programs

numerous ways to achieve the credit point. There is no set procedure to achieve a given credit. Therefore, authors choose to use prescriptive pathways to achieve credits and eliminate performance pathways. Table 4.8 reports nine instances where performance pathway is eliminated from this book. Similarly, any other pathway that can be followed to achieve any credit using any other assessment tool (refer Section 4.3.6) is also eliminated.

In the "IEQ11.2 Surface Illuminance" credit, there is a separate pathway that is referred to as the "IEQ11.2C Residential Spaces." This pathway is specifically meant for residential units. Therefore, this credit is eliminated

TABLE 4.8 Credits eliminated.

Key criteria	Credit	Credit point	Eliminated credit options
Management (M)	M5.1 Environmental building Performance	1 point	M5.1.1B Certified Operational Performance Ratings
	M8 Operational Waste	1 point	M8A Performance Pathway
Indoor environment quality (IEQ)	IEQ11.1.2 Glare reduction	1 point	IEQ11.1.2C Performance Method
	IEQ11.2 Surface Illuminance	1 point	IEQ11.2B Performance Method IEQ11.2C Residential Spaces
	IEQ12.1 Daylight	2 points	IEQ12.1B Compliance using the Daylight Factor (Performance method) IEQ12.1B Compliance using Daylight Autonomy (Performance method)
	IEQ13.1.2 Carpets	Up to 2 points (from IEQ13.1 paints, adhesives, sealants, and carpets)	IEQ13.1.2B Laboratory testing
Energy (E)	E15 Greenhouse Gas Emissions	Up to 20 points	E15B GHG Emissions Reduction —NatHERS E15C GHG Emissions Reduction—BASIX E15D GHG Emissions Reduction—NABERS Energy Commitment Agreement E15E GHG Emissions Reduction—Modeled Performance
	E16 Peak Electricity Demand Reduction	Up to 2 points	E16B Modeled Performance Pathways: Reference Building
Transport (T)	T17 Sustainable Transport	Up to 10 points	T17A Performance Pathway
Water (W)	W18 Potable Water	Up to 12 points	W18A Prescriptive Pathway

Continued

TABLE 4.8 Credits eliminated—cont'd

Key criteria	Credit	Credit point	Eliminated credit options
Materials (Mat)	Mat19A Life Cycle Impacts	Up to 7 points	Mat19A Performance Pathway—Life Cycle Assessment
	Mat19B Life Cycle Impacts: Steel	Up to 2 points	19B.2B Reduced Mass of Steel Reinforcement
Materials (Mat	Mat22 Construction and Demolition Waste	1 point	Mat22A Reduction of Construction and demolition Waste—Percentage Benchmark

because the model is developed for office buildings. In the "IEQ13.1.2 Carpets" credit, two pathways are available (refer Section 4.3.3). From these two available pathways, the "IEQ13.1.2B Laboratory testing" is eliminated because authors use product certification to support compliance with the requirement.

The building considered is a concrete-framed structure. Therefore, the "19B.2B Reduced Mass of Steel Reinforcement" credit, which focuses on steel-framed structures is eliminated. As illustrated in Section 4.3.3, authors use waste minimization strategies, and, therefore, "Mat22A Reduction of Construction and demolition Waste—Percentage Benchmark" credit is eliminated.

In certain instances, the elimination of these credits does not impact the achievement of credit points. This is because even though a particular credit is eliminated, there is another pathway to obtain that particular credit point. However, in certain other instances, when a credit is eliminated certain credit points are also eliminated. For example, "Greenhouse Gas Emissions" can obtain up to 20 credit points by using the "E15E GHG Emissions Reduction—Modelled Performance" pathway. This is a performance pathway, and therefore it is eliminated. With the elimination of the performance pathway and the certification of other assessment tools, the "E15A GHG Emissions Reduction—Prescriptive pathway" remains the only pathway to gain credit points. However, this pathway gains only up to five credit points. Therefore, under the "Greenhouse Gas Emissions" sub-criteria, a maximum of 15 credit points is excluded due to the elimination of credit points.

In the "IEQ11.2 Surface Illuminance" credit, both the "IEQ11.2B Performance Method" and "IEQ11.2C Residential Spaces" are eliminated; additionally, instead of these two credits, the "IEQ11.2A Prescriptive Method" is used. The number of credit points in the prescriptive method is similar to that of the performance pathway. Therefore, in this credit, the selected credit points selected remain unchanged.

This book considered 63 credits for the life-cycle cost model, with a total credit point value of 74. However, if a building requires a 6-star rating, then it would need more than 75 credits in total. Therefore, the model cannot satisfy these boundaries. As a result, this model only considers up to five-star rating for commercial buildings. Table 4.9 reports a list of credits used for the model after eliminating all the credits.

The credits used in the model (refer to Table 4.9) have various inter-dependencies. The life-cycle cost calculation and the cost model captured these inter-dependencies in cost calculation and optimum credit selections.

4.4 Dependencies among credits

When all these credits are considered together, many inter-dependencies are found among credits. Although there are credits that focus on same type of building services, they are classified under different key criteria. Furthermore, there are certain credits that can be collectively considered for cost calculations and bundled together for optimum selection of credits.

Mechanical services (for HVAC), sanitary and plumbing services, and electrical services are key building services in any commercial building. Therefore, these separate systems are considered separately while estimating the life-cycle cost. Besides this, there are different credits that support each other to meet the requirement. Fig. 4.3 illustrates all the relationships among credits and the different services under which each credit can be categorized.

In Fig. 4.3, different building services are depicted in different colors. There are three main services discussed in these credits, namely, electrical services, mechanical services, and sanitary and plumbing services. Additionally, Fig. 4.3 shows the dependencies between credits by linking the credits with each other.

There are many credits attributed to mechanical services (refer Fig. 4.3). According to Fig. 4.3, these credits spread across three key criteria, namely IEQ, water, and emissions. Most of the credits related to mechanical services represent the IEQ key criteria, and all the credits representing the IEQ key criteria share a direct relationship with air quality and thermal comfort. The "W18B.3 Heat Rejection" credit is classified under the water key criterion. However, this credit is directly related to the mechanical services. To achieve this credit, the building must either be naturally-ventilated or have an HVAC system that does not use water for heat rejection. Therefore, this credit can be considered when calculating the life-cycle costs for mechanical services. Similarly, "Em28 Legionella

TABLE 4.9 List of credits used for the model.

Key criteria	Sub-criteria	Credits	Credit point(s)
Management (M)	Green Star Accredited Professional	M1.0 Accredited Professional	1 point
	Commissioning and Tuning	M2.0 Environmental Performance Targets	Required
		M2.1 Services and Maintainability Review	1 point
		M2.2 Building Commissioning	1 point
		M2.3 Building Systems Tuning	1 point
		M2.4 Independent Commissioning Agent	1 point
	Adaptation and Resilience	M3.0 Implementation of a Climate Adaptation Plan	2 points
	Building Information	M4.1 Building Operations and Maintenance Information	1 point
		M4.2 Building User Information	1 point
	Commitment to Performance	M5.1 Environmental Building Performance	1 point
		M5.2 End of Life Waste Performance	1 point
	Metering and Monitoring	M6.0 Metering	Required
		M6.1 Monitoring Systems	1 point
	Construction Environmental Management	M7.0 Environmental Management Plan	Required
		M7.1 Formalized Environmental Management System	1 point
	Operational Waste	M8B Prescriptive pathway: Facilities	1 point

Indoor environment quality (IEQ)	Indoor Air Quality	IEQ9.1 Ventilation System Attributes	1 point
		IEQ9.2 Provision of Outdoor Air	2 points
		IEQ9.3 Exhaust or Elimination of Pollutants	1 point
	Acoustic Comfort	IEQ10.1 Internal Noise Levels	1 point
		IEQ10.2 Reverberation	1 point
		IEQ10.3 Acoustic Separation	1 point
	Lighting comfort	IEQ11.0 Minimum Lighting Comfort	Required
		IEQ11.1 General illuminance and glare Reduction	1 point
		IEQ11.2 Surface Illuminance	1 point
		IEQ11.3 Localized Lighting Control	1 point
	Visual Comfort	IEQ12.0 Glare reduction	Required
		IEQ12.1 Daylight	2 points
		IEQ12.2 Views	1 point
	Indoor Pollutants	IEQ13.1 Paints, Adhesives, Sealants, and Carpets	1 point
		IEQ13.2 Engineered Wood Products	1 point
	Thermal Comfort	IEQ14.1 Thermal Comfort	1 point
		IEQ14.2 Advanced Thermal Comfort	1 point
Energy (E)	Green House Gas Emissions	E15A GHG Emissions Reduction—Prescriptive Pathway	5 points
	Peak Electricity Demand Reduction	E16A Prescriptive Pathway: On-site Energy Generation	1 point

TABLE 4.9 List of credits used for the model—cont'd

Key criteria	Sub-criteria	Credits	Credit point(s)
Transport (T)	Sustainable Transport	T17B.1 Prescriptive Pathway: Access by public transport	3 points
		T17B.2 Prescriptive Pathway: Reduced Car Parking Provisions	1 point
		T17B.3 Prescriptive Pathway: Low Emission Vehicle Infrastructure	1 point
		T17B.4 Prescriptive Pathway: Active Transport Facilities	1 point
		T17B.5 Prescriptive Pathway: Walkable Neighborhood	1 point
Water (W)	Potable Water	W18B.1 Prescriptive Pathway: Sanitary Fixture Efficiency	1 point
		W18B.2 Prescriptive Pathway: Rainwater Reuse	1 point
		W18B.3 Prescriptive Pathway: Heat Rejection	2 points
		W18B.4 Prescriptive Pathway: Landscape Irrigation	1 point
		W18B.5 Prescriptive Pathway: Fire System Test Water	1 point
Materials (Mat)	Life Cycle Impacts	Mat19B Prescriptive Pathway—Life Cycle Impacts	5 points
	Responsible Building Material	Mat 20.1 Structural Reinforcing Steel	1 point
		Mat 20.2 Timber Products	1 point
		Mat 20.3 Permanent, Formwork, Pipes, Flooring, Blinds, and Cables	1 point
	Sustainable Products	Mat 21 Product Transparency and Sustainability	3 points
	Construction and Demolition Waste	Mat 22 Reduction of Construction and Demolition Waste	1 point

Land use and ecology (L)	Ecological Value	L23.0 Endangered, Threatened, or Vulnerable species	Required
		L23.1 Ecological Value	3 points
	Sustainable Sites	L24.0 Conditional Requirement	Required
		L24.1 Reuse of Land	1 point
		L24.2 Contamination and hazardous Material	1 point
	Heat Island Effect	L25 Heat Island Effect Reduction	1 point
Emissions (Em)	Storm water	Em26.1 Reduced Peak Discharge	1 point
		Em26.2 Reduced Pollution Targets	1 point
	Light Pollution	Em27.0 Light Pollution to Neighboring Bodies	Required
		Em27.1 Light Pollution to Night Sky	1 point
	Microbial Control	Em28 Legionella Impacts from Cooling Systems	1 point
	Refrigerant Impacts	Em29 Refrigerant Impact	1 point

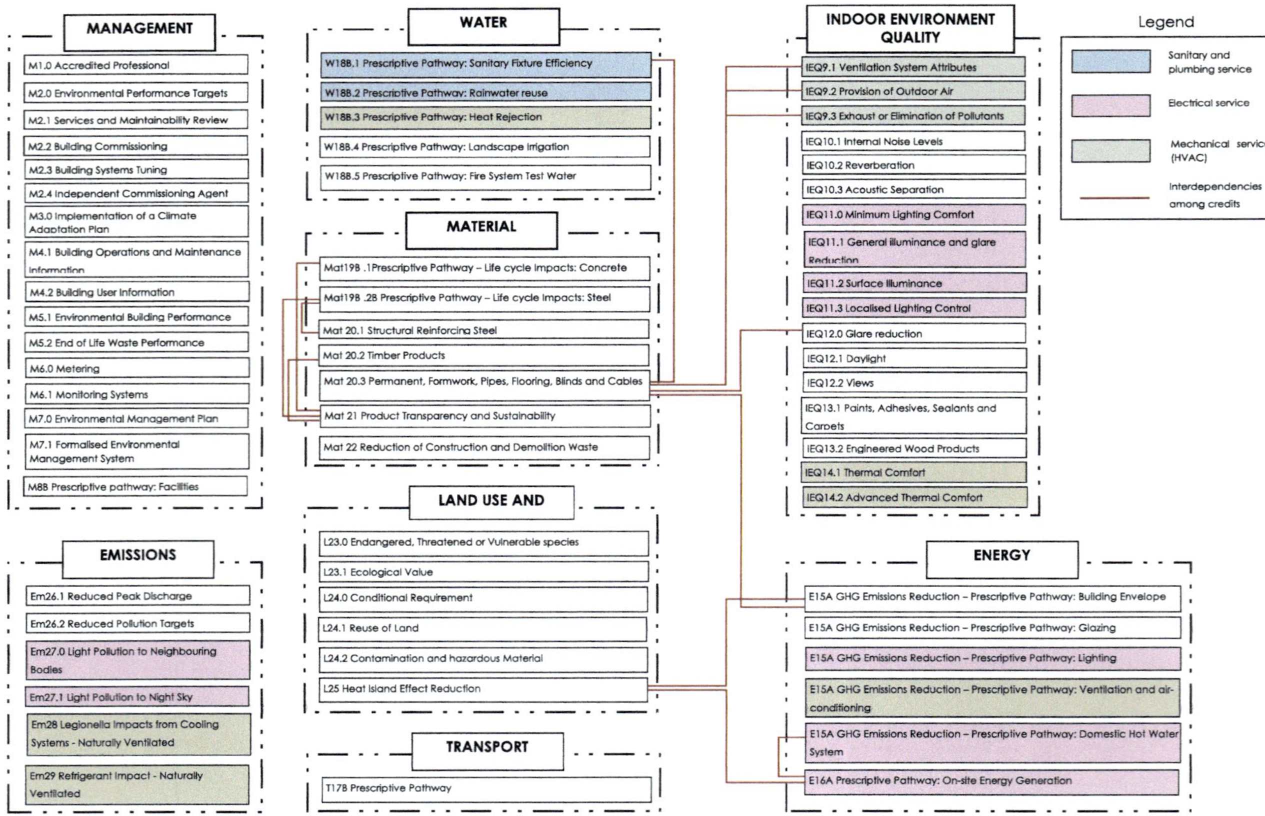

FIG. 4.3 Dependencies among credits.

Impacts from Cooling Systems" and "Em29 Refrigerant Impact" credits strongly contribute toward the mechanical services, although these credits are listed under the emissions key criterion. The users can achieve the "Em28 Legionella Impacts from Cooling Systems" credit, based on the type of the HVAC system of the building. This credit can be achieved directly if the building is naturally-ventilated or if it has a waterless heat rejection system. Therefore, this credit is considered when developing the life-cycle cost for mechanical services.

According to the Green Star Design and As-Built version 1.1, any mechanical equipment utilized to air-condition a space is considered a "refrigerant equipment" for the purpose of this credit (Green Building Council of Australia, 2015). The main aim of this credit is to minimize the impact of refrigerants leaking into the environment. Therefore, this credit is included in the life-cycle cost calculations of mechanical services. Additionally, the achievement of the "E15A GHG Emissions Reduction—ventilation and air conditioning" credit also directly depends on the HVAC system of the building.

Electrical services include many credit points in the rating system (refer Fig. 4.3). Similar to the mechanical services, most of the credits considered under the electrical services were classified under the IEQ key criterion. These credits are directly related to the lighting of the building, illuminance, and glare. The "E15A GHG Emissions Reduction—Lighting" credit under the energy key criterion is also directly related to the electrical services. Additionally, the "E16A Prescriptive Pathway: On-site Energy Generation" credit also has a significant impact on the electrical services. The existence of a renewable energy generation source in a building leads to significant energy cost savings during the building's life-cycle. Finally, the "Em27.0 Light Pollution to Neighboring Bodies" and "Em27.1 Light Pollution to Night Sky" credits are also considered as part of the electrical services during life-cycle cost calculations.

Sanitary and plumbing services include two credits that are classified under the water key criterion, namely, "W18B.1 Prescriptive Pathway—Sanitary Fixture Efficiency" and "W18B.1 Prescriptive Pathway—Rainwater Reuse." Both these credits have the same plumbing and water services requirements. Therefore, both the credits are considered together.

There are certain links between credits. These credits are collectively considered for selecting the optimum solutions in the life-cycle cost model. According to Fig. 4.3, the material key criterion has many interdependent credits within the key criteria and many intra-dependent credits among various key criteria. The "Mat 21 Product Transparency and Sustainability" credit requires the materials used for the building to meet transparency and sustainability requirements; this can be achieved by reusing and recycling materials, using environmental product declarations (EPDs), using third party certificates, or using stewardship programs.

Therefore, this credit can be linked to other credits under the same key criterion. For example, the "Mat19B Prescriptive Pathway—Life Cycle Impacts" credit can be achieved by reducing the use of building material by partially replacing cement content with supplementary cementitious material (SCM) material and reducing the use of steel. Therefore, these two credits can be inter-linked. Similarly, "Mat 20.2 Timber Products" credit requires either the reuse of timber or the use of timber with a certification. Therefore, again there is a link between these two credits because if "Mat 20.2 Timber Products" is achieved, then there would also be a possibility to achieve "Mat 21 Product Transparency and Sustainability" by satisfying the requirement of reusing material.

The "Mat 20.3 Permanent, Formwork, Pipes, Flooring, Blinds and Cables" credit shares several links with other credits. According to the Green Star Design and As-Built version 1.1, all pipes, flooring, blinds, and cables either should not contain polyvinyl chloride (PVC) or meet the best practice guidelines for PVC. Usually, plumbing services use PVC pipes. Therefore, in order to achieve the "Mat 20.3 Permanent, Formwork, Pipes, Flooring, Blinds and Cables" credit, PVC pipes should not be used. Furthermore, this is also applicable to electrical services. Similarly, concerning blinds, if the user aims to achieve the "IEQ12.0 Glare Reduction" credit, then blinds containing PVC components should not be used. Additionally, the user that aims to meet the requirements of the "E15A GHG Emissions Reduction—Building Envelope" credit should not consider the use PVC products when deciding materials for enveloping the building. The aforementioned examples show that all these credits are inter-linked. All these credits must be considered together when selecting optimum solutions and for life-cycle cost calculation.

Under the energy criterion, there is a direct link between the "E16A Prescriptive Pathway: On-site Energy Generation" and "E15A GHG Emissions Reduction—Domestic Hot Water" credits. According to the Green Star Design and As-Built version 1.1, the "E15A GHG Emissions Reduction—Domestic Hot Water" credit can be achieved by demonstrating that the hot water system is powered by a renewable energy source. Therefore, if a building focuses on using a renewable energy source, then the users can achieve both the aforementioned credits collectively. Similarly, when calculating the areas for the "L25 Heat Island Effect Reduction," the user must consider the roofing material and hardscaping elements, including photo voltaic (PV) panels. However, the type of roofing material has a direct effect on the "E15A GHG Emissions Reduction—Building Envelope" credit, and the use of PV panels depends on the "E16A Prescriptive Pathway: On-site Energy Generation" credit. Therefore, these two credits share a link with the "L25 Heat Island Effect Reduction."

The life-cycle cost calculations consider all these inter-dependencies and links. Furthermore, these dependencies are also considered when selecting optimum solutions from the life-cycle cost model.

4.5 Summary

This chapter discussed the Green Star Design and As-Built version 1.1credits and various classifications under these credits. Initially, this chapter defined the terminology used to identify different levels of information provided in the rating tools, such as key criteria, sub-criteria, credits, and credit points. Subsequently, this chapter classified credits into various types and identified conditional credits, additional credits, different pathways to achieve credits, various options for credits, credits based on the type of building, credits linked to other assessment tools, credits with choices, and the credits eliminated from this book when developing the cost model. After conducting a thorough analysis of the credits, this chapter reported a table listing all the credits considered in this book. Finally, this chapter illustrated inter-dependencies among various credits. These inter-dependencies are very critical and significant for life-cycle cost calculations and the selection of optimum solutions.

Reference

Green Building Council of Australia. (2015). *Green Star—Design and As Built Version 1.1*. Retrieved from https://www.gbca.org.au/green-star/green-star-design-as-built/the-rating-tool/.

CHAPTER

5

Life-cycle cost calculation for each key criterion

5.1 Introduction

This chapter focuses on the life-cycle cost calculations for this study. Life-cycle cost data is one of the significant inputs in this study because the proposed life-cycle cost model uses these cost data to select the optimum solutions. Cost calculations for each of the key credit criteria are illustrated under separate headings in this chapter.

5.2 Methodology for life-cycle cost calculation

Life-cycle cost calculation signifies a major portion of this book. As illustrated in Section 3.3, there are various cost components in life-cycle cost calculation which occurs in various stages in green building life-cycle. Therefore, this section identifies the standards used in life-cycle cost calculation and illustrates various cost calculations and respective equations separately.

According to the Australian National Audit Office (2001), the process of life-cycle costing fundamentally involves assessing costs that arise from an asset over its lifespan, and evaluating alternatives that have an impact on this cost of ownership. Further, there are five main phases which trigger different types of costs in the lifespan of an asset. These phases are design, purchase and construction, operations, maintenance, and development and disposal (Australian National Audit Office, 2001). Based on the length of the lifespan of a building, the design, purchase, and construction costs may represent the initial stages of the construction process. Therefore, "initial costs" represent the current market prices. However, since development costs represent major refurbishments made to buildings, they are not considered in this study. Operational and maintenance costs occur throughout the building life-cycle. Finally, the disposal costs

https://doi.org/10.1016/B978-0-12-820062-9.00005-0

represent the costs incurred in demolishing the building. Apart from the initial costs, all other costs occur in different phases of the building life-cycle. Therefore, these costs need to be discounted back to present values.

The life-cycle cost calculation follows the ISO 15686-5:2017, namely, "Building and construction assets—service life planning—Part 5: Life-cycle costing standard" as a guideline (International Organisation for Standardization [ISO], 2017). The net present value (NPV) technique is used to calculate life-cycle costs. Eq. (5.1) shows the formula for the NPV calculation.

$$\text{NPV calculation}: \boldsymbol{NPV}(\boldsymbol{i},\boldsymbol{N}) = \sum_{t=0}^{N} \frac{\boldsymbol{R_t}}{(\boldsymbol{1}+\boldsymbol{i})^{t}} \tag{5.1}$$

where i denotes the discount rate, t denotes the time of cash flow, R_t denotes the net cash flow, and N is the total number of periods. The discount rate is established considering the time value of money and the associated risk. The minimum attractive rate of return is commonly used as the discount rate (Dell'Isola & Kirk, 2003). The interest rate on a 25-year treasury bond in Australia is 3.25% per annum (Australian Government, 2016). Furthermore, the return on assets for a non-residential construction firm is approximately 3.30% (Deloitte Access Economics, 2016). Therefore, the discount rate is taken as 3.25% for this calculation. However, this rate changes for each user, because the associated risk differs from person to person. Therefore, in order to calculate the life-cycle cost, a user should identify the associated discount rate. Thus, in the life-cycle cost model, the discount rate is a variable input.

According to International Organisation for Standardization [ISO] (2017), life-cycle costing in construction commences from the planning stage and ends with the disposal stage. Therefore, this study follows a "gate to grave" analysis on life-cycle costs for green office buildings. The time period for this life-cycle cost calculation is 60 years, as required by the Green Building Council of Australia (2015). Further, all the costs are normalized to 1 m^2 of gross floor area (GFA) of the building.

There are different types of costs that occur within the life-cycle of a building. Therefore, it is necessary to identify the costs that should be included in the life-cycle cost calculation. The International Organisation for Standardization [ISO] (2017) illustrated that the life-cycle cost includes construction costs, operation costs, maintenance costs, end-of-life costs, and environmental costs. Externalities and social benefits are not considered in life-cycle costs, as these costs fall under whole-life costing, and not life-cycle costing (International Organisation for Standardization [ISO], 2017). In this study, the construction costs are termed as "initial cost" because these costs are incurred in the initial stages of the life-cycle, and also include management costs related to construction, as well.

5.2.1 Determination of initial cost

The basic initial costs are collected for the six main central business districts (CBDs) in Australia, namely, Adelaide, Brisbane, Hobart, Melbourne, Perth, and Sydney. All the prices are excluding Goods and Services Tax (GST) and profit. The initial cost is developed based on first principles using current market prices, and includes all the material, labor, and plant costs. As an example, the initial material cost for external walls included the costs of insulation material, bricks or blocks as applicable, cladding, insulation membrane, and wall lining. This cost occurs in the current time period, and therefore, this value is directly included in the life-cycle cost, without any discounting.

5.2.2 Determination of maintenance, replacement, and other costs

Each component of a building has different maintenance requirements, depending on the materials used, exposure to the external environment, and system requirements. These maintenance requirements are based on maintenance manuals and guidance from maintenance engineers on specific products and materials. Systems such as air conditioning require regular maintenance, on an annual basis. Further, components such as external walls require maintenance, at specific intervals, depending on the material used.

As an example, most external wall options require repointing joints at regular intervals. Further, there are minor repairs as well. Floor structures require general inspection and minor repairs. Usually, such different wall and floor options require proactive maintenance to prevent costly repairs and full replacement. The maintenance requirements are further developed based on the detailed analysis provided by Dell'Isola and Kirk (2003), and Stanford (2010).

Apart from maintenance, this study considered replacement of different material, as well. As an example, usually, external walls have a life span longer than 60 years. However, timber flooring requires replacement in 50 years (Dell'Isola & Kirk, 2003). In such a scenario, the timber flooring is replaced in the fiftieth year of the life-cycle and the cost is discounted to the present value, using Eq. (5.1).

If maintenance occurs annually over the 60 years lifespan of a building, the annual cost should be discounted 60 times. To avoid excessive calculations, the present value of annuity (PVA) for a period of 60 years is calculated using Eq. (5.2).

$$\text{PVA calculation}: PVA = R_m \times \left(\frac{1-(1+i)^{-N}}{i} \right) \tag{5.2}$$

where i denotes the discount rate, R_m denotes the annual maintenance cost, and N is the total number of periods.

On the lines of maintenance costs, there are others, such as utility costs that occur within the operational stages of a building. As an example, if the electrical system installed within the building is considered, there are regular electricity costs that the user needs to pay to the electricity provider. Further, due to certain changes in the system, the user might get certain cost savings as well. All these costs and savings occur within the operational stages of a building and that needs to be reflected in the life-cycle cost calculation. Therefore, these costs represent all the "other costs" in this calculation.

5.2.3 Determination of disposal and other related costs

This study considered disposing the building and its components at the end of the life-cycle. However, there are re-usable materials in these structures, such as bricks. Therefore, for all re-usable material, an additional cost is added for the extra care required during demolition. As an example, the life-cycle cost calculation includes the cost of preparing timber framing for reuse. The debris is assumed to be transported 15 km away from the site. Disposal of these items occur at the end of the life-cycle. The life-cycle cost calculations include all these related costs and discount them to the present value using Eq. (5.1).

The life-cycle cost calculation is carried out for each credit point. This enables the developers to identify the life-cycle costs for achieving each credit, and to derive the minimum life-cycle cost when provided with necessary inputs and constraints. The life-cycle cost calculation differs for each credit. Certain credits have an initial cost alone. For example, the "M1.0 Accredited Professional" credit has an initial cost during the design and construction phase. However, credits such as "M6.0 Metering" and "IEQ9.2 Provision of Outdoor Air" incur maintenance costs. In such cases, the NPV calculation is used to discount the future cash flow. As illustrated earlier while developing the cost model, the discount rate changes for different users. Therefore, while calculating life-cycle costs, initial costs, maintenance replacement and other costs, and disposal costs are calculated separately. The life-cycle cost calculation for each criterion is given below in Eq. (5.3).

$$\text{Life}-\text{cycle cost for key criterion}: C_{sum} = \sum_{n=C(\alpha)}^{C(\beta)} x_n + \sum_{n=C(\alpha)}^{C(\beta)} y_n + \sum_{n=C(\alpha)}^{C(\beta)} z_n \quad (5.3)$$

where x denotes the initial cost per square metre for each credit point of a given criterion, while y denotes the present values of maintenance replacement and other costs, z denotes present value of demolition costs, and C_{sum} denotes the sum of all the credit points of the given criteria. The "x" component in this equation does not require any discounting to arrive at the present value. However, "y" and "z" components in Eq. (5.3) require the discount rate in the NPV calculations. Further, the "y" component has cash outflows occurring regularly, and Eq. (5.3) is used for the calculation. Therefore, in the life-cycle cost model, the "y" and the "z" components change constantly based on the users' discount rates. Similarly, the total life-cycle cost can be calculated for all the remaining criteria.

Costs vary significantly across the main CBDs of Australia. In this manner, a user can select the main CBD and then use regional indices to obtain accurate cost data. Finally, the location-adjusted total life-cycle cost calculation is given in Eq. (5.4).

$$\text{Location}-\text{adjusted total life}-\text{cycle cost}: LCC_{total} = \left[M_{sum} + IEQ_{sum} + T_{sum} + E_{sum} + Mat_{sum} + W_{sum} + L_{sum} + Em_{sum}\right] \times RI \quad (5.4)$$

where LCC_{total} denotes the location-adjusted total life-cycle cost, while RI denotes the regional indices, which are obtained from Rawlinson's cost guide (Rawlinsons, 2017).

While calculating the life-cycle cost for this study, sensible assumptions were made when necessary. Further, although the technique used is illustrated in this section, there are methods that were carried out to capture different cost data and some credits were combined for cost calculations. Therefore, this chapter illustrates the life-cycle cost calculation process for each credit in each key criterion, separately.

5.3 Introduction to life-cycle cost calculation for credits

In green buildings, there are certain aspects which are discussed throughout the operational period. Most green buildings rating tools mainly focus on these criteria, such as indoor environmental quality (IEQ), energy, water, and operational waste (Building Construction Authority, 2013; Building Research Establishment Environment Assessment Method, 2014; Green Building Council of Australia, 2015; United States Green Building Council, 2014). Most of the benefits of green buildings such as better human conditions, efficiency in energy and water, life-cycle cost savings, and lower negative impact to the environment are derived through the operational phase, with effective operations of these aspects.

As mentioned in Section 5.2, all the life-cycle cost calculations are developed based on the first principles of cost estimations. Therefore, prior to the life-cycle cost calculations, the authors identified the cost components included in the life-cycle cost. According to Eq. (5.3), there are three components of costs, namely, the initial cost per square meter for each credit point of a given criterion, the present values of maintenance, replacement, and other costs, and the present value of demolition costs. Each of the credits in this study have been analyzed thoroughly to identify the costs included. After analyzing all the credits in Green Star Design and As-Built version 1.1, Illankoon, Tam, Le, and Shen (2016), illustrated that out of all the credits with a contribution to life-cycle cost, 3% of the credit points were from management key criteria, 22% were from the energy criteria, 8% came from the IEQ criteria, 12% came from both, material and water criteria, 5% were the emission key criteria, and 1% came from land used and ecology key criteria. However, there are conditional credits (refer Section 4.3.1) without credit points that are not included in these percentage calculations (Illankoon et al., 2016). Therefore, in summary a minimum of 63% of credit points contributes directly to life-cycle cost calculations. However, since life-cycle cost calculations are rarely carried out in the initial stages of decision making, the life-cycle cost impact is completely ignored. As an example, there are many options for green buildings with lower initial costs. However, the life-cycle costs attached to those options are significant. Similarly, there are options with higher initial costs and lower operational costs. In both of these situations, if the decisions are taken based on the initial costs, there is a higher possibility that the selected option does not represent the optimum solutions. This illustrates the significance of carrying out life-cycle cost calculation at the initial stages of the decision making process for the green buildings which aims to obtain green building certification. To obtain green building certification, it is necessary to achieve the credits representing various key criteria as illustrated in Chapter 4. Therefore, this study presents life-cycle cost calculations for each credit in latter stages (refer Table 5.3).

Further, the impact of the costs other than the initial cost is significant for certain credits and the impact is minimal for certain other credits. Therefore, it is necessary to identify the different cost components in all the credits in this study. The life-cycle cost calculations require certain parameters to be determined. The life-cycle cost fluctuates based on changes in the prices of materials. To establish the life-cycle cost accurately, cost and time-related variables must be established. It is also necessary to predict the influence of these uncertain variables in the absence of accurate data by using certain techniques, such as sensitivity analysis (Wong, Perera, & Eames, 2010). Therefore, the main uncertain parameters are identified and sensitivity analysis is carried out.

The NPV calculation and the maintenance cost significantly depend on the discount rates being considered. Therefore, it is necessary to calculate the sensitivity of discount rates towards the final life-cycle cost calculation, if the maintenance, replacement, and other costs contribute to a significant portion of the life-cycle cost. Therefore, for each key criterion, there is a

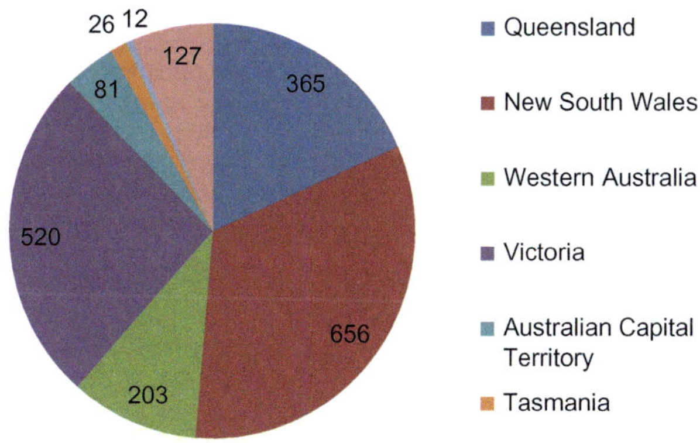

FIG. 5.1 State breakdown of certified green buildings in Australia. *Data Source: Green Building Council of Australia. (2018). About GBCA. Retrieved from: http://www.gbca.org.au/about/.*

separate sensitivity analysis calculated to account for the uncertainty of this study. The prices of material also changes over time. However, the proposed model is developed to regularly update the changes in cost, which is explained in detail in the following chapters. Therefore, sensitivity to the changes in prices are not considered in this chapter.

As illustrated in Section 5.2, this study calculated life-cycle cost for six main CBDs in Australia. According to Green Building Council of Australia (2018), most of the certified green buildings represents the state of New South Wales. Fig. 5.1 below illustrates the state breakdown of 1990 certified green buildings.

According to Fig. 5.1, New South Wales, Victoria and Queensland representing Sydney, Melbourne and Brisbane CBDs respectively, have the first three green building markets with higher number of green certifications. Therefore, finding necessary green material and suppliers were comparatively easier compared to other CBDs. This is reflected in cost data as well.

5.4 Management key criterion

The management key criterion can have up to a maximum of 14 credit points according to Green Star Design and As-Built version 1.1. The main aim of this key criterion is to encourage and reward the adoption of practices and processes that enable and support best practice sustainability outcomes throughout the different phases of project design, construction, and its ongoing operations (Green Building Council of Australia, 2015). In this key criterion, there are many credits that occur within the duration of the design and construction of the building. The costs incurred within

TABLE 5.1 Cost components of credits in the management key criterion.

Credit	Initial cost	Maintenance, replacement and other costs	Demolition cost
M1.0 Accredited Professional	✓	–	–
M2.0 Environmental Performance Targets	✓	–	–
M2.1 Services and Maintainability Review	✓	–	–
M2.2 Building Commissioning	✓	–	–
M2.3 Building Systems Tuning	✓	✓	–
M2.4 Independent Commissioning Agent	✓	–	–
M3.0 Implementation of a Climate Adaptation Plan	✓	–	–
M4.1 Building Operations and Maintenance information	✓	–	–
M4.2 Building User Information	✓	–	–
M5.1 Environmental Building Performance	✓	✓	–
M5.2 End of Life Waste Performance	✓	–	✓
M6.0 Metering	✓	✓	–
M6.1 Monitoring Systems	✓	✓	–
M7.0 Environmental Management Plan	✓	–	–
M7.1 Formalised Environmental Management System	✓	–	–
M8B Prescriptive pathway: Facilities	✓	✓	✓

that period are taken together as the initial cost, because these costs are included in the cost calculation at the time of cost estimation of the project. Table 5.1 reports the different cost components of management credits.

The "M2.3 Building Systems Tuning" credit requires including quarterly adjustments and measurements for the first 12 months after occupation. Therefore, this credit has a cost component that occurs within the operational stages of the building, and yet the cost impact is very minimal (refer Table 5.1). Similarly, according to Table 5.1, there are only very few credits with maintenance and demolition cost components. Further, even though there is a maintenance cost component in this credit, its contribution towards the total life-cycle cost is very minimal.

Most credits in management key criterion require higher levels of costs for documentation and professional fees, especially in the design and construction stages. The professional fees are obtained from the industry itself.

5.4.1 Life-cycle cost for management credits

Most credits in the management criterion included consultancy fees for various professionals. Further, many credits required professional services, and these are directly obtained from consultants. Table 5.2 reports the equations used for life-cycle cost calculations.

Based on these formulae and other calculations, life-cycle cost is calculated and reported in Table 5.3.

All the credits in the management criterion have lower life-cycle costs except for very few, which have considerably higher life-cycle costs. There are three conditional credits in this criterion. While selecting the optimum solutions, these credits must also be considered. As an example, the "M6.1 Monitoring system" credit has a lower life-cycle cost. However, if this credit is achieved, it is a must to have achieved the "M6.0 Metering" credit beforehand. Therefore, the total life-cycle cost needs to include the cost relation to the conditional credits as well.

5.4.2 Optimum solutions for management credits

Fig. 5.2 below reports all the optimum solutions for management credits. There are many options available with lower life-cycle costs. However, based on Fig. 5.2, the optimum solutions are:

- "M1.0 Accredited Professional"
- "M2.1 Services and Maintainability Review"
- "M4.1 Building Operations and Maintenance Information"
- "M4.2 Building User Information"
- "M5.1 Environmental Building Performance"
- "M5.2 End of Life Waste Performance"

TABLE 5.2 Equations used for management credits.

$Cost\ of\ consultancy\ fee = Number(Nr)\ of\ hours\ required\ weekly \times Hourly\ professional\ rate \times 52$	Eq. (5.5): Consultancy fee calculations
$Life-cycle\ cost\ for\ monitoring = [Nr\ of\ hours\ annually \times Nr\ of\ professionals \times Hourly\ rate\ per\ professional] \times PV$	Eq. (5.6): Life-cycle cost for monitoring
$\text{Life} - \text{cycle cost for operational waste management system (OWM)} = Initial\ cost\ for\ OWM\ system + (Nr\ of\ hours\ for\ annual\ maintenance \times Nr\ of\ labourers \times Hourly\ labour\ rate) \times PV + NPV\ for\ cleaning\ garbage\ chute\ every\ year + NPV\ of\ replacement\ cost\ after\ 20\ years + NPV\ of\ cost\ of\ demolition$	Eq. (5.7): Life-cycle cost for operational waste management system

TABLE 5.3 Life-cycle cost for management credits.

Discount rate—3.25%							
		Life-cycle cost [Australian Dollar (AUD)/m^2]					
Credits	Credit point	Adelaide	Brisbane	Hobart	Melbourne	Perth	Sydney
M1.0 Accredited Professional	1	4.13	3.45	3.41	3.75	4.18	4.17
M2.0 Environmental Performance Targets	Required	4.56	3.81	3.79	4.16	4.63	4.63
M2.1 Services and Maintainability Review	1	8.74	7.31	7.26	7.97	8.88	8.87
M2.2 Building Commissioning	1	21.28	17.80	17.68	19.40	21.62	21.59
M2.3 Building Systems Tuning	1	19.00	15.90	15.79	17.33	19.31	19.28
M2.4 Independent Commissioning Agent	1	25.78	21.56	21.33	23.44	26.11	26.11
M3.0 Implementation of a Climate Adaptation Plan	2	11.40	9.54	9.47	10.40	11.58	11.57
M4.1 Building Operations and Maintenance Information	1	7.98	6.68	6.63	7.28	8.11	8.91
M4.2 Building User Information	1	7.60	6.36	6.31	6.93	7.72	7.71
M5.1 Environmental Building Performance	1	5.70	4.77	4.74	5.20	5.79	5.78
M5.2 End of Life Waste Performance	1	3.80	3.18	3.16	3.47	3.86	3.86
M6.0 Metering	Required	4.18	3.50	3.47	3.81	4.25	4.24
M6.1 Monitoring Systems	1	6.08	5.09	5.05	5.54	6.18	6.17
M7.0 Environmental Management Plan	Required	4.26	3.56	3.54	3.88	4.32	4.32
M7.1 Formalized Environmental Management System	1	6.84	5.72	5.68	6.24	6.95	6.94
M8B Prescriptive pathway: Facilities	1	9.30	7.78	7.78	8.48	9.30	9.30

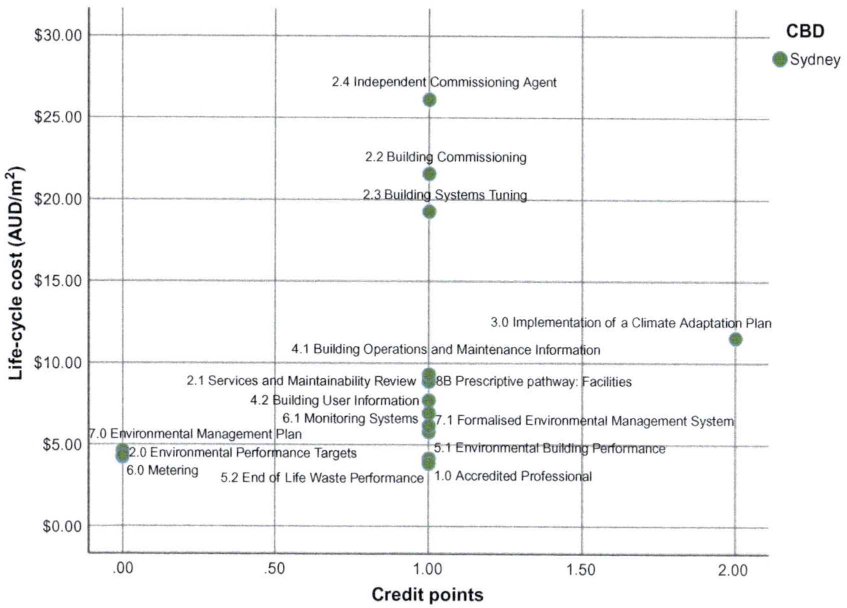

FIG. 5.2 Optimum solutions for management credits.

Apart from these options, credits such as "M6.1 Monitoring Systems" and "M7.1 Formalized Environmental Management System" have lower costs within the optimum range (refer Fig. 5.2). However, they require conditional credits, and therefore, these are not considered in the optimum solutions.

There are certain changes to the life-cycle cost when compared with different CBDs. Therefore, Fig. 5.3 illustrates the changes in life-cycle cost in the main CBDs.

According to Fig. 5.3, all the CBDs have similar life-cycle costs, except for credits such as "M2.2 Building Commissioning," "M2.3 Building Systems Tuning" and "M2.4 Independent Commissioning Agent." In these credits, the life-cycle cost is considerably higher in Sydney. The main reason for this increment is the higher costs for professional services.

5.5 IEQ key criterion

As the name suggests, IEQ refers to the internal environment quality, or, the quality of the environment within the building, especially focusing on the occupants. According to the Green Building Council of Australia (2015), IEQ key criterion aims to encourage and reward initiatives that enhance the comfort and well-being of occupants. Most of the credits included in this key criterion include all the three cost components of the

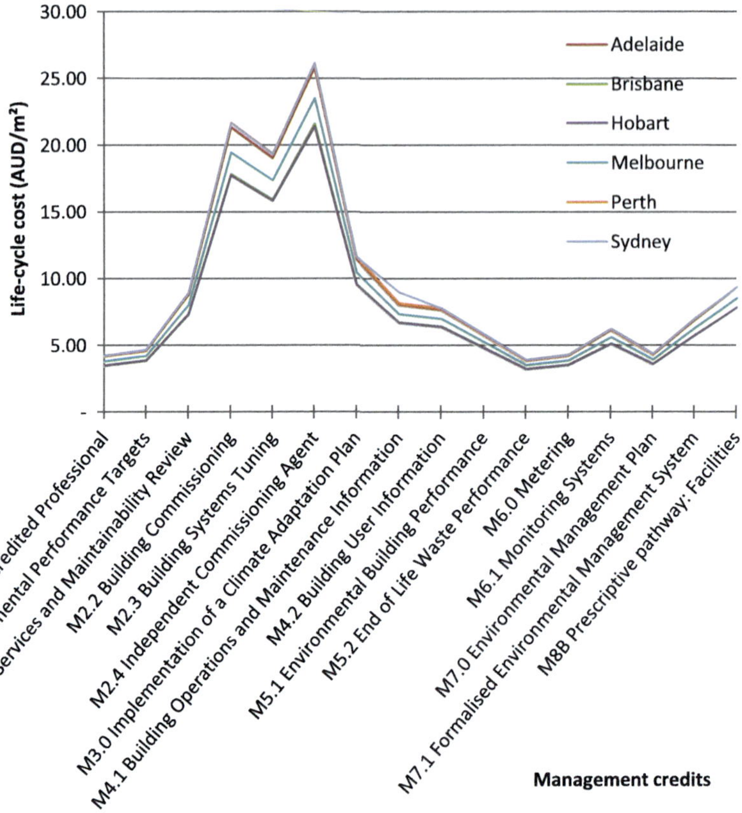

FIG. 5.3 Life-cycle cost for management credits in six CBDs.

life-cycle cost. Table 5.4 reports all the cost components included in the IEQ credits, separately. According to Table 5.4, most of the credits contribute to all the cost components of the life-cycle cost.

In the IEQ key criterion, credits such as "IEQ9.1 Ventilation System Attributes," "IEQ9.2 Provision of Outdoor Air," "IEQ14.1 Thermal Comfort" and "IEQ14.2 Advanced Thermal Comfort" have different options to achieve the credit points, depending on the type of ventilation system in use. As an example, if a building is naturally ventilated, the standards and systems considered for that type of ventilation are completely different from those considered for a building with mechanical ventilation. Therefore, while calculating the life-cycle cost, these two types are calculated separately, and these costs have significant differences in their values. Further, according to Section 4.4, there is a dependency on "W18B.3 Heat Rejection," whose credit points are awarded when there is no water used for heat rejection. Therefore, a mechanically ventilated system is further divided into two, based on the type of heat rejection

TABLE 5.4 Cost components of credits in IEQ key criterion.

Sub criteria	Credit	Initial cost	Maintenance, replacement and other costs	Demolition cost
Indoor Air Quality	IEQ9.1 Ventilation System Attributes	✓	✓	✓
	IEQ9.2 Provision of Outdoor Air	✓	✓	✓
	IEQ9.3 Exhaust or Elimination of Pollutants	✓	✓	✓
Acoustic Comfort	IEQ10.1 Internal Noise Levels	✓	✓	–
	IEQ10.2 Reverberation	✓	✓	–
	IEQ10.3 Acoustic Separation	✓	✓	–
Lighting Comfort	IEQ11.0 Minimum Lighting Comfort	✓	✓	–
	IEQ11.1 General illuminance and glare Reduction	✓	✓	–
	IEQ11.2 Surface Illuminance	✓	✓	–
	IEQ11.3 Localised Lighting Control	✓	✓	–
Visual Comfort	IEQ12.0 Glare reduction	✓	✓	✓
	IEQ12.1 Daylight	✓	–	–
	IEQ12.2 Views	✓	–	–
Indoor Pollutants	IEQ13.1 Paints, Adhesives, Sealants and Carpets	✓	✓	✓
	IEQ13.2 Engineered Wood Products	✓	✓	✓
Thermal Comfort	IEQ14.1 Thermal Comfort	✓	✓	✓
	IEQ14.2 Advanced Thermal Comfort	✓	✓	✓

method in use. Finally, life-cycle cost calculations are carried out for naturally ventilated systems, mechanically ventilated systems using water for heat rejection, and mechanically ventilated systems using air for heat rejection, separately.

5.5.1 Life-cycle cost calculation for IEQ credits

Life-cycle cost is calculated for each of the credits in the IEQ criterion. The calculation formulae are illustrated in Table 5.5.

By following the given formulae, life-cycle cost can be calculated. An example of the life-cycle cost calculation considering a discount rate 3.25% is given in Table 5.6.

As per Green Star Design and As-Built version 1.1, a building needs to be evaluated by an acoustic consultant to achieve acoustic comfort credits. Therefore, Eq. (5.11) included professional fees for an acoustic consultant. The cost of demolition is not included separately in Eq. (5.11). The main reason for this is that sound insulation material will be demolished along with the building structure and allocating separately for demolition of sound insulation is not needed, as the cost is very minimal.

In Eq. (5.10), 26% of energy savings are allocated to the lighting system, because, according to the Department of Climate Change and Energy Efficiency (2012), in Australian commercial office buildings, the lighting system consumes 26% of the total energy consumption. Therefore, energy savings are also apportioned based on the energy consumption of the building. Similarly, in Eq. (5.18), only 43% of energy savings are considered due to the proportion of energy consumed by the HVAC system Department of Climate Change and Energy Efficiency (2012).

5.5.2 Optimum solutions for IEQ credits

There are various credits which can be selected from the IEQ key criteria. However, the optimum solutions are the credits with the lowest life-cycle cost and higher number of credit points. Therefore, Fig. 5.4 presents the optimum solution considering the IEQ credits. The life-cycle cost is calculated for Sydney and the discount rate is 3.25%.

Fig. 5.4 is based on data from Sydney. However, costs change across the CBDs. Therefore, Fig. 5.5 presents the changes in life-cycle costs in various CBDs, when the discount rate is 3.25%.

According to Fig. 5.4, the optimum credits in the IEQ key criterion are follows:

- "IEQ12.1 Daylight"—Obtained through a better design
- "IEQ9.2 Provision of Outdoor Air—Naturally Ventilated"—Rather than mechanically ventilated water cooled system and air cooled system, natural ventilated system is more suited in terms of life-cycle costs

TABLE 5.5 Equations used for IEQ credits.

$\text{Life}-\text{cycle cost for providing air to the building} = \textit{Initial cost for providing outdoor air} + (\textit{Annual maintenance}) \times \textit{PVa} + \textit{NPV for cleaning duct work every 5 years} + \textit{NPVfor replacing ductwork and other components every 25 years} + \textit{NPV of cost of demolition}$	Eq. (5.8): Life-cycle cost for providing air to the building
$\textit{Annual maintenance cost} = \textit{Nr of hours per session} \times \textit{Nr of Air Conditioning mechanics} \times \textit{Hourly rate per Air Conditioning mechanic}$	Eq. (5.9): Annual maintenance cost for provisioning air
$\text{Life}-\text{cycle cost for exhaust air ducts} = \textit{Initial cost for providing exhaust air duct} + (\textit{Nr of hours for annual maintenance} \times \textit{Nr of labourers} \times \textit{Hourly labour rate}) \times \textit{PV} + \textit{NPV for cleaning exhaust ducts every 5 years} + \textit{NPV of replacement cost after 30 years} + \textit{NPV of cost of demolition}$	Eq. (5.10): Life-cycle cost for providing exhaust ducts to the building
$\text{Life}-\text{cycle cost for acoustic comfort} = \textit{Consultancy fee for the acoustic consultant} + \textit{Initial cost of sound insulation} + \textit{NPV for annual maintenance for every 5 years}$	Eq. (5.11): Life-cycle cost for providing acoustic comfort
$\text{Life}-\text{cycle cost for internal lighting} = \textit{Initial cost for light fittings, fixtures and lighting points} + (\textit{Annual maintenance} \times \textit{PVA}) + \textit{Replacement cost of lamps every 3 years} + \textit{NPV of annual energy saving} \times 0.26 + \textit{NPV of cost of demolition}$	Eq. (5.12): Life-cycle cost for providing lighting comfort
$\textit{Annual maintenance cost} = [(\textit{Nr of hours for annual maintenance} \times \textit{Nr of labourers} \times \textit{Hourly labour rate}) + (\textit{Nr of hours for annual maintenance} \times \textit{Nr of electricians} \times \textit{Hourly rate for Electrician})] + (\textit{Nr of hours for monthly visual inspection and cleaning for lighting fixtures and controls} \times \textit{Hourly labour rate} \times 12)$	Eq. (5.13): Annual maintenance cost for electrical system

Continued

TABLE 5.5 Equations used for IEQ credits—contsd

$\text{Life} - \text{cycle cost for internal blinds}$ $= \textit{Initial cost for internal blinds including fixing}$ $+ \textit{NPV of maintenance cost every 2 years including replacement if required}$ $+ \textit{NPV of cost of demolition}$	Eq. (5.14): Life-cycle cost for providing internal blinds for visual comfort
$\text{Life} - \text{cycle cost for paints}$ $= \textit{Initial cost for painting} + (\textit{Annual maintenance}) \times \textit{PVA}$ $+ \textit{NPV for repainting every 7 years}$	Eq. (5.15): Life-cycle cost for providing internal painting
$\text{Life} - \text{cycle cost for carpets}$ $= \textit{Initial cost for carpets} + \textit{NPV of maintenance cost every 2 years}$ $+ \textit{NPV of replacement cost every 15 years}$ $+ \textit{NPV of cost of demolition}$	Eq. (5.16): Life-cycle cost for carpets
$\text{Life} - \text{cycle cost for engineered wood products}$ $= \textit{Initial cost for interior doors including fixing}$ $+ (\textit{Annual maintenance cost}) \times \textit{PVA} + \textit{NPV for demolition}$	Eq. (5.17): Life-cycle cost for engineered wood products
$\text{Life} - \text{cycle cost for thermal comfort}$ $= \textit{Initial cost for HVAC system excluding ductwork}$ $+ (\textit{Annual maintenance} \times \textit{PVA}) + \textit{Replacement costs every 3 years}$ $+ \textit{NPV of annual energy saving} \times 0.43$ $\times + \textit{NPV of cost of demolition}$	Eq. (5.18): Life-cycle cost for HVAC system
$\textit{Annual maintenance cost}$ $= [(\textit{Nr of hours for daily inspections} \times \textit{Nr of labourers}$ $\times \textit{Hourly labour rate}) \times 365]$ $+ (\textit{Nr of hours for annual maintenance}$ $\times \textit{Nr of Air Conditioning technicians}$ $\times \textit{Hourly rate for Air Conditioning technician})$ $+ [(\textit{Nr of hours for quarterly maintenance}$ $\times \textit{Nr of Air Conditioning technicians}$ $\times \textit{Hourly rate for Air Conditioning technician} \times 4)]$	Eq. (5.19): Annual maintenance cost for HVAC system

TABLE 5.6 Life-cycle cost for IEQ credits.

Discount rate—3.25% Ventilation type—Mechanically ventilated using water cooled heat rejection							
		Life-cycle cost (AUD/m^2)					
Credits	**Credit point**	**Adelaide**	**Brisbane**	**Hobart**	**Melbourne**	**Perth**	**Sydney**
IEQ9.1 Ventilation System Attributes	1	15.41	12.89	12.80	15.41	15.65	15.63
IEQ 9.2 Provision of Outdoor Air—Mechanically Ventilated—Water Cooled system	2	124.21	124.38	124.32	124.66	123.83	124.28
IEQ 9.3 Exhaust or Elimination of Pollutants	1	20.76	17.37	17.25	18.93	21.09	21.06
IEQ 10.1 Internal Noise Levels	1	6.05	5.06	5.02	5.51	6.14	6.13
IEQ 10.2 Reverberation	1	7.50	6.27	6.23	6.84	7.62	7.61
IEQ 10.3 Acoustic Separation	1	8.22	6.88	6.83	7.50	8.35	8.34
IEQ 11.0 Minimum Lighting Comfort	Required	45.70	44.45	44.95	42.46	48.44	45.20
IEQ 11.1 General illuminance and glare Reduction	1	60.93	59.27	59.93	56.61	64.58	60.27
IEQ 11.2 Surface Illuminance	1	64.74	62.97	63.68	60.15	68.62	64.03
IEQ 11.3 Localised Lighting Control	1	76.16	74.09	74.92	70.76	80.73	75.33
IEQ 12.0 Glare reduction	Required	33.49	28.01	27.82	30.53	34.02	33.97
IEQ 12.1 Daylight	2	15.63	13.07	12.98	14.25	15.88	15.85
IEQ 12.2 Views	1	6.70	5.60	5.56	6.11	6.80	6.79
IEQ 13.1 Paints, Adhesives, Sealants and Carpets	1	194.25	177.40	162.54	178.61	189.59	181.66
13.2 Engineered Wood Products	1	28.66	23.97	23.71	26.06	29.03	29.00
IEQ 14.1 Thermal Comfort—Mechanically Ventilated—Water Cooled system	1	155.26	155.48	155.39	155.83	154.79	155.35
IEQ 14.2 Advanced Thermal Comfort—Mechanically—Water Cooled system ventilated	1	124.21	124.38	124.32	124.66	123.83	124.28

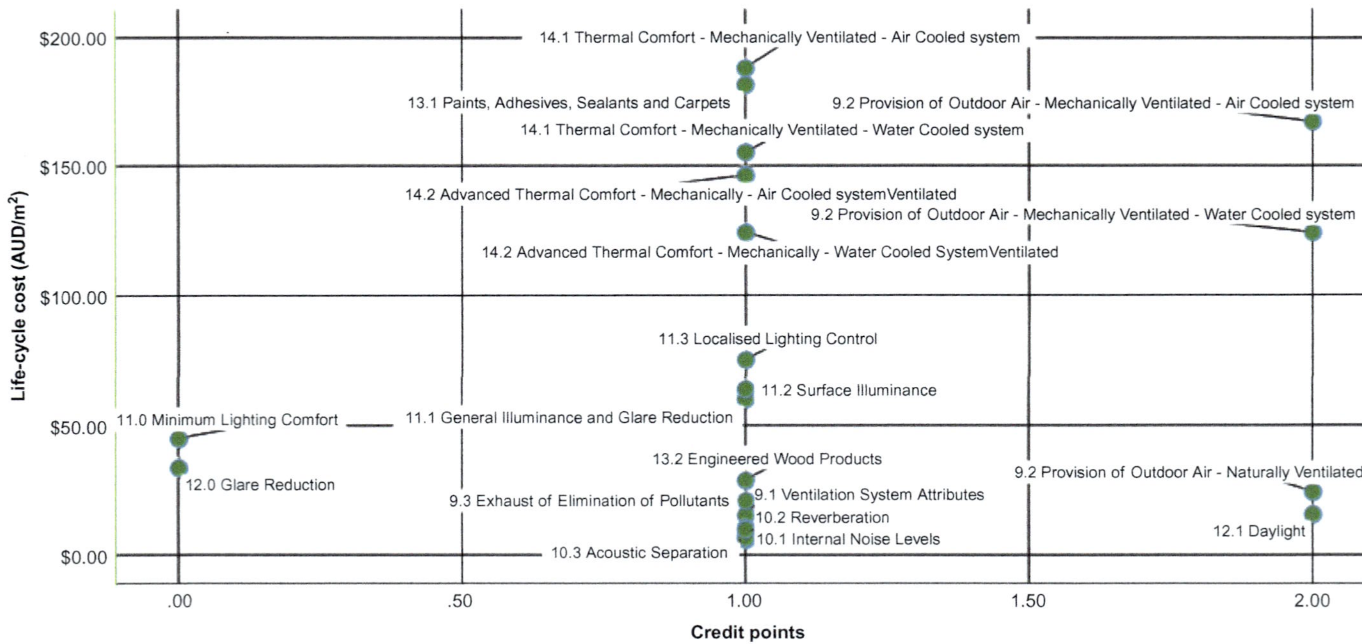

FIG. 5.4 Optimum solutions for the IEQ credits.

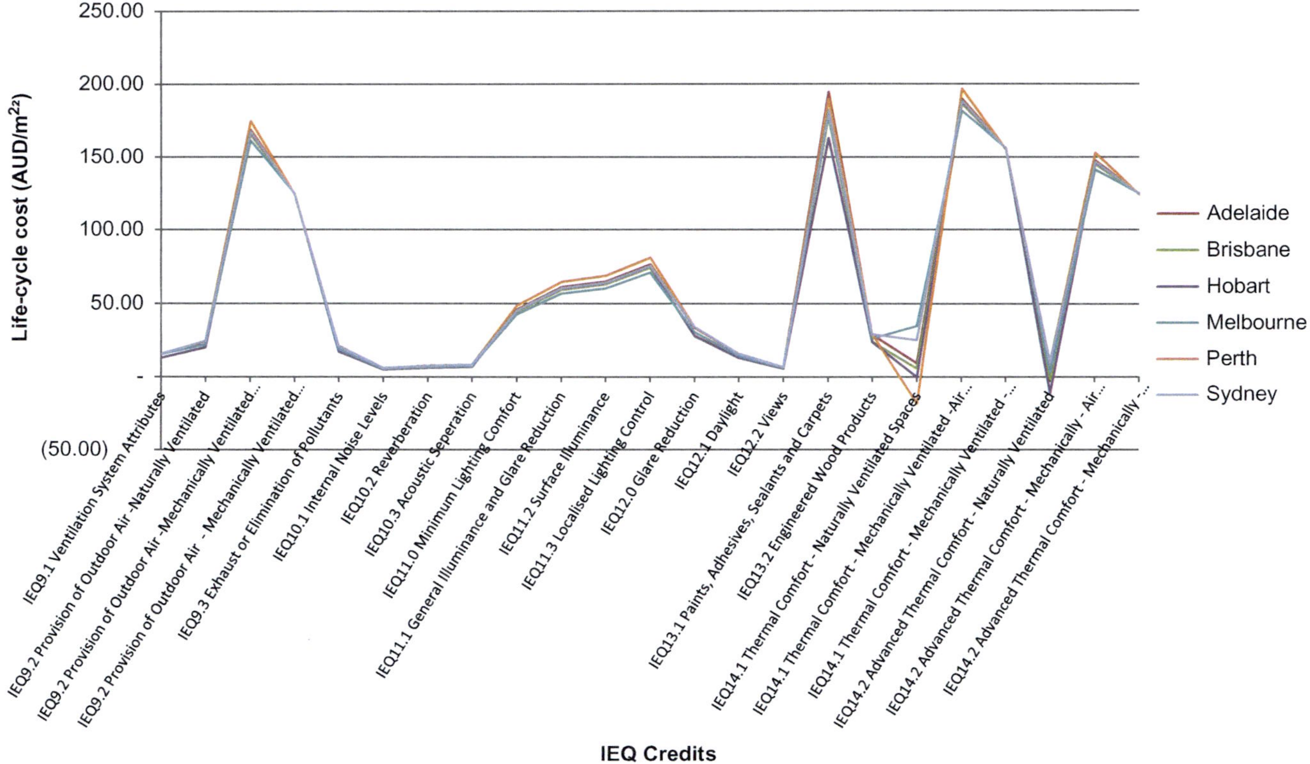

FIG. 5.5 Life-cycle cost for the IEQ credits in six CBDs.

- "IEQ10.1 Internal Noise Levels"
- "IEQ10.2 Reverberation"
- "IEQ10.3 Acoustic Separation"

While considering Fig. 5.5, there are only slight changes in the life-cycle costs in various CBDs, except for credits "IEQ14.1 Thermal Comfort—Naturally ventilated System" and "IEQ14.2 Advanced Thermal Comfort—Naturally Ventilated System." The changes are due to the changes in energy costs among the CBDs. When there are higher energy costs, the life-cycle cost is significantly reduced because of the life-cycle savings. As an example, according to Fig. 5.5, Melbourne has the lowest life-cycle cost due to "IEQ14.1 Thermal Comfort—Naturally ventilated System" because of lower electricity rates.

"IEQ9.2 Provision of Outdoor Air—Naturally Ventilated System" has various inter-dependencies, and this can be further optimized by considering various credits in other key criteria (refer Section 4.3). While using the proposed life-cycle cost model, these inter-dependencies can be selected further for optimum solutions.

There can be changes in the solution based on changes in the discount rates. Therefore, it is necessary to carry out a sensitivity analysis to identify whether there are significant changes in life-cycle costs leading to changes in the optimum solutions, or not.

5.5.3 Sensitivity analysis for the IEQ credits

Most of the credits in the IEQ key criterion have significant maintenance costs. Further, certain credits have cost savings throughout the life-cycle of the building. As an example, due to the sufficient light fittings used in the building there are electricity savings. Therefore, this is included in the operational stages of the building. These savings are also reflected in the maintenance, replacement, and other cost components of the life-cycle costs. However, these costs directly depend on the discount rates used in the calculation. Therefore, Fig. 5.6 reports the sensitivity analysis of each of these credits, to explain the changes in the discount rate.

Fig. 5.6 illustrates different life-cycle cost figures for the IEQ credits for an office building located in Sydney with mechanically ventilated air conditioning using water for heat rejection. There are four discount rates used to illustrate the changes in the life-cycle cost (refer Fig. 5.6). There is a reduction in life-cycle cost when the discount rate increases, and this reduction is significant when the contribution of costs during the operational phase is higher in the life-cycle cost calculation. This is significantly evident in credits relating to the lighting system of a building. As an example, credits such as "IEQ11.0 Minimum Lighting Comfort," "IEQ11.1 General

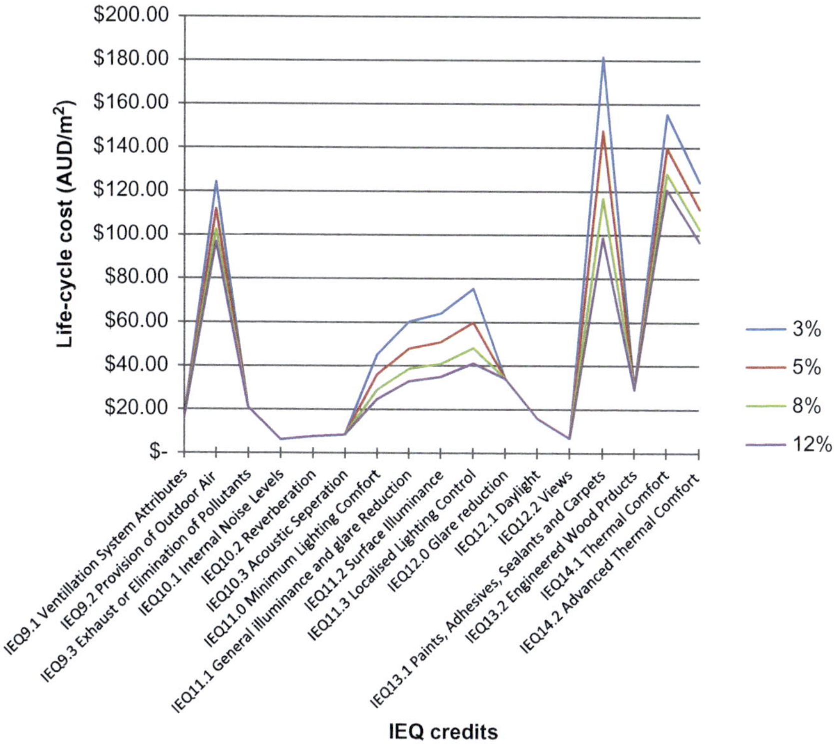

FIG. 5.6 Sensitivity analysis for IEQ credits.

Illuminance and glare Reduction," and "IEQ11.2 Surface Illuminance" have significant maintenance, replacement, and other cost proportions. Further, they have considerable costs for electricity throughout the life-cycle of the building. Therefore, with the changes in the discount rates, the changes in the life-cycle cost are significantly evident. However, credits such as "IEQ10.1 Internal Noise Levels," "IEQ10.2 Reverberation," and "IEQ10.3 Acoustic Separation" have minimum costs during the operational phases. Therefore, the maintenance, replacement, and other cost component is at a minimum, and thus, changes to the discount rate are also minimum (refer Fig. 5.6).

The sensitivity of the life-cycle cost to the changes in discount rate is significant in certain IEQ credits. As an example, if "IEQ12.0 Glare reduction" and "IEQ11.2 Surface Illuminance" credits are considered when the discount rate is 3%, there is significant difference in the life-cycle costs among these two credits (refer Fig. 5.6). However, if the discount rate is 12%, there is a slight difference between the two credits (refer Fig. 5.6). This shows the importance of using the most suited and applicable discount rates to achieve an accurate and consistent optimum selection of credits. This sensitivity analysis further illustrates that most of the IEQ credits are sensitive to the changes in the discount rate.

5.6 Energy key criterion

The main aim of the energy key criterion is to reward projects that are designed and constructed to reduce their overall operational energy consumption to below that of a comparable standard-practice building (Green Building Council of Australia, 2015). After the elimination of certain credits, (refer Section 4.3.8), the key criterion has two main credits. Table 5.7 reports the cost components contributed to these credits.

All the credits in the energy criterion make significant contributions to costs that are incurred during the operational phase of the building. There are five credit elements that have been separately illustrated to achieve "E15A GHG Emissions Reduction—Prescriptive Pathway." These credit elements included building envelopes, glazing, lighting, ventilation and air conditioning, and domestic hot water systems. All these credit elements had inter-dependencies with other credits as well (refer Section 4.4). The building envelope credit focused on roof, floor, and wall insulation, according to the given conditions. There are many options available for these requirements, and considering the optimum solution, focusing on higher thermal resistance and lower life-cycle cost, authors selected the best material for the model (Illankoon, Tam, & Le, 2018). Apart from that, glazing, lighting, and air conditioning elements need to fulfill specific requirements. The domestic hot water system needs to be powered by a renewable energy source. Therefore, this credit is bundled with the other credit, namely, "E16A Prescriptive Pathway: On-site Energy Generation."

The "E16A Prescriptive Pathway: On-site Energy Generation" credit requires PV panels as renewable on-site energy generation sources. Due to the incentives and other benefits given to energy generation through PV panels, there are considerable changes in the life-cycle costs among

TABLE 5.7 Cost components of credits in energy key criterion.

Credit	Initial cost	Maintenance, replacement and other costs	Demolition cost
E15A GHG Emissions Reduction—Prescriptive Pathway	✓	✓	✓
E16A Prescriptive Pathway: On-site Energy Generation	✓	✓	✓

different CBDs. However, in all the CBDs, there were reported savings in the life-cycle costs (Tam, Le, Zeng, Wang, & Illankoon, 2017).

5.6.1 Life-cycle cost for energy credits

Life-cycle cost calculations for energy credits included many formulae. These are given in Table 5.8.

An example of the life-cycle cost calculation for Sydney considering a discount rate 3.25% is given in Table 5.9.

According to Table 5.9, there five options are discussed to achieve "E15GHG Emissions Reduction." Further, depending on the type of ventilation system in the building, there are three options available. Further, while using PV panels for on-site energy generation, there are savings on the life-cycle costs.

5.6.2 Optimum solutions for energy credits

There are many options to be selected in the energy criterion. Considering the lowest life-cycle costs and higher number of credit points, the optimum solutions can be selected. Therefore, Fig. 5.7 presents

TABLE 5.8 Equations used for energy credits.

$\text{Life}-\text{cycle cost for building envelope} = \textit{Initial cost for building envelope including insulation} + \textit{Maintenance cost} + \textit{NPV of cost of demolition}$	Eq. (5.20): Life-cycle cost for building envelope
$\textit{Maintenance cost} = \textit{NPV of (Nr of hours for maintenance per square meter every } 10 \textit{ years} \times \textit{Hourly labour rate)}$	Eq. (5.21): Maintenance cost for building envelope
$\text{Life}-\text{cycle cost for glazing} = \textit{Initial cost for glazing} + \textit{(Nr of hours for annual maintenance} \times \textit{Nr of labourers} \times \textit{Hourly labour rate)} \times \textit{PVA} + \textit{NPV of replacement cost} + \textit{NPV of cost of demolition}$	Eq. (5.22): Life-cycle cost for glazing
$\text{Life}-\text{cycle cost for domestic hot water (DHW) system} = \textit{Initial cost of DHW system} + \textit{NPV for maintenance for every } 5 \textit{ years} + \textit{NPV for annual energy saving} + \textit{NPV of cost of demolition}$	Eq. (5.23): Life-cycle cost for domestic hot water system
$\text{Life}-\text{cycle cost for PV panels} = \textit{Initial cost for PV panels including fixing} + \textit{(Nr of hours of annual maintenance} \times \textit{Hourly rate for technicians)} \times \textit{PVA} + \textit{NPV of cost of demolition}$	Eq. (5.24): Life-cycle cost for PV panels

TABLE 5.9 Life-cycle cost for energy credits.

Discount rate—3.25%							
		Life-cycle cost (AUD/m^2)					
Credits	**Credit Point**	**Adelaide**	**Brisbane**	**Hobart**	**Melbourne**	**Perth**	**Sydney**
E15A GHG Emissions Reduction—Building Envelope	1	103.81	114.41	115.85	109.96	119.62	108.68
E15A GHG Emissions Reduction—Glazing	1	146.77	144.37	141.57	141.57	146.77	146.77
E15A GHG Emissions Reduction—Lighting	1	76.16	74.09	74.92	70.76	80.73	75.33
E15A GHG Emissions Reduction-Mechanically Ventilated—Air Cooled System	1	105.27	103.54	104.23	100.78	109.07	104.58
E15A GHG Emissions Reduction-Mechanically Ventilated—Water Cooled System	1	93.16	93.29	93.24	93.50	92.87	93.21
E15A GHG Emissions Reduction—Naturally Ventilated	1	1.46	1.20	1.18	1.30	1.56	1.48
E15A GHG Emissions Reduction—Domestic Hot Water System	1	16.85	17.74	17.38	19.16	14.40	17.21
E16A Prescriptive Pathway: On-site Energy Generation	1	– 18.05	– 15.09	– 14.93	– 16.41	– 18.28	– 18.26

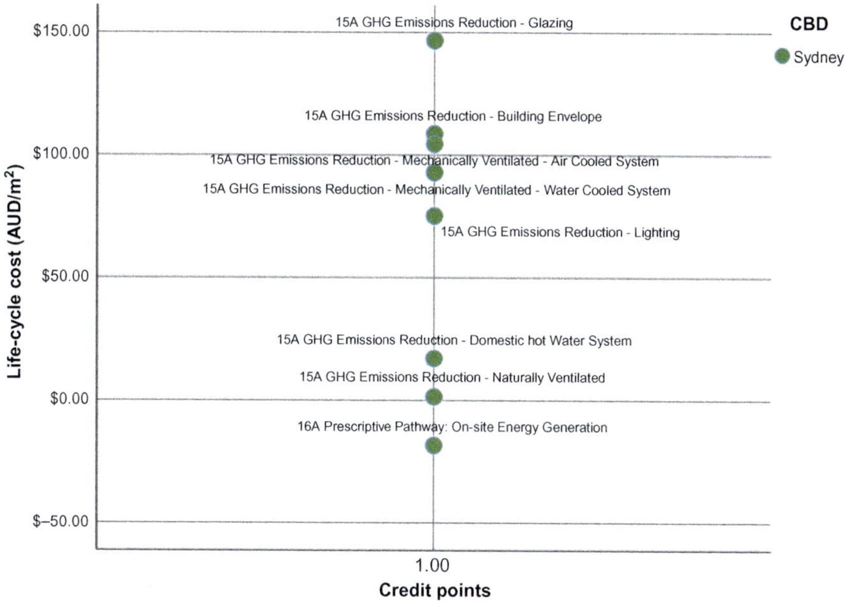

FIG. 5.7 Optimum solutions for energy credits.

the optimum solutions considering energy credits. The life-cycle cost is calculated for Sydney and the discount rate is 3.25%. The optimum solutions are straightforward, considering Fig. 5.7.

According to Fig. 5.7, the optimum credits in the energy key criterion are follows:

- "E16A Prescriptive Pathway: On-site Energy Generation"—obtained by using PV panels. This credits achieves energy saving
- "E15A GHG Emissions Reduction—Naturally Ventilated"—Rather than mechanically ventilated water cooled system and air cooled system, the natural ventilated system is more suited in terms of life-cycle costs
- "E15A GHG Emissions Reduction—Domestic hot Water System"
- "E15A GHG Emissions Reduction—Lighting"

Fig. 5.7 is based on data for Sydney. However, costs change across the CBDs. Therefore, Fig. 5.8 presents the changes in life-cycle costs in various CBDs, when the discount rate is 3.25%.

According to Fig. 5.8, there are only minimal changes in life-cycle costs across the CBDs. However, these minimal changes are due to changes in price levels, in energy costs, and in labor rates. All these life-cycle costs are based on the discount rate of 3.25%. Energy credits have maintenance and replacement costs involved. Therefore, it is necessary to identify the sensitivity to the changes in the discount rate.

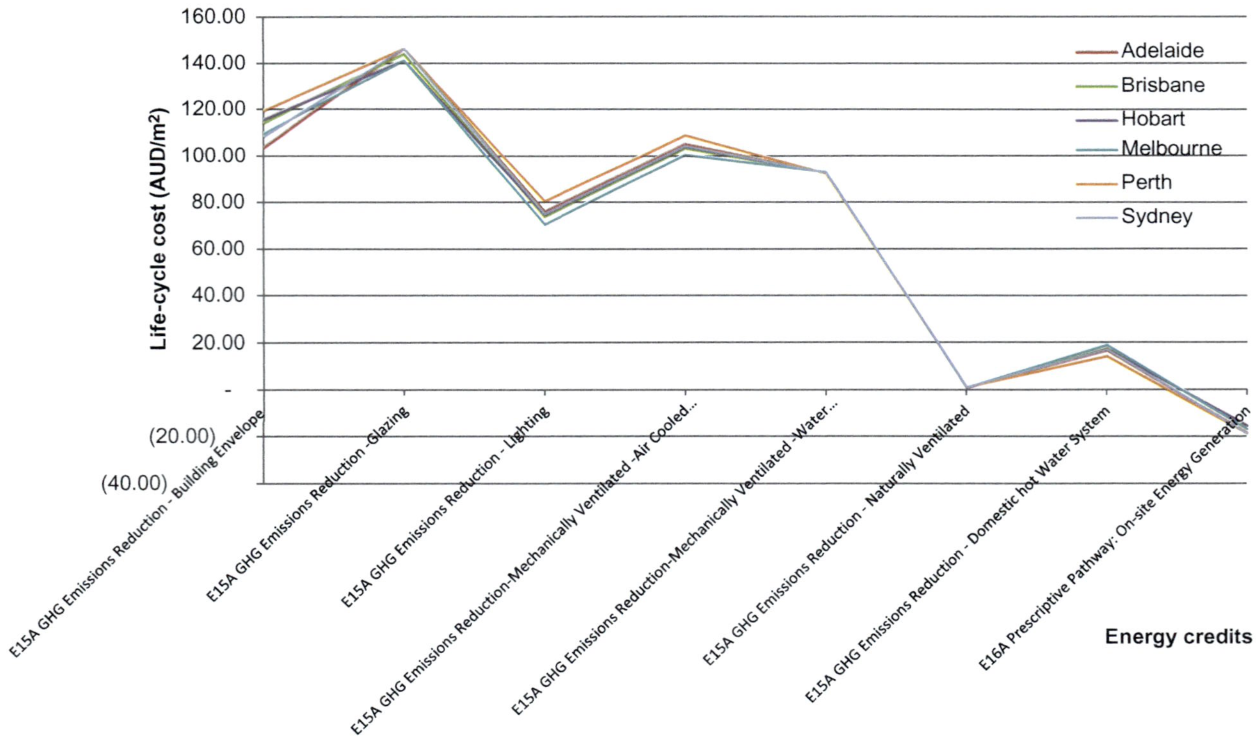

FIG. 5.8 Life-cycle costs for energy credits in six CBDs.

5.6.3 Sensitivity analysis for energy credits

Similar to the IEQ credits, energy credits also have maintenance, disposal, replacement, and other cost components in the life-cycle cost calculation. Therefore, Fig. 5.9 illustrates the sensitivity to the life-cycle cost and its impact on the discount rate. The life-cycle costs are for an office building in Sydney and the discount rate changes from 3%, to 5%, to 8%, and to 12%.

According to Fig. 5.9, there is a major impact on the life-cycle cost, especially for "E16A Prescriptive Pathway: On-site Energy Generation." For lower discount rates, there is a cost saving for using the PV panels for energy generation. However, the cost saving is canceled if the discount rate is higher. Further, a majority of other credits also have substantial impacts on discount rate, and thus, on the life-cycle cost. Therefore, similar to the IEQ criterion, while using the proposed model, users need to be cautious and input the most suited discount rate to reflect the risk and cost of capital.

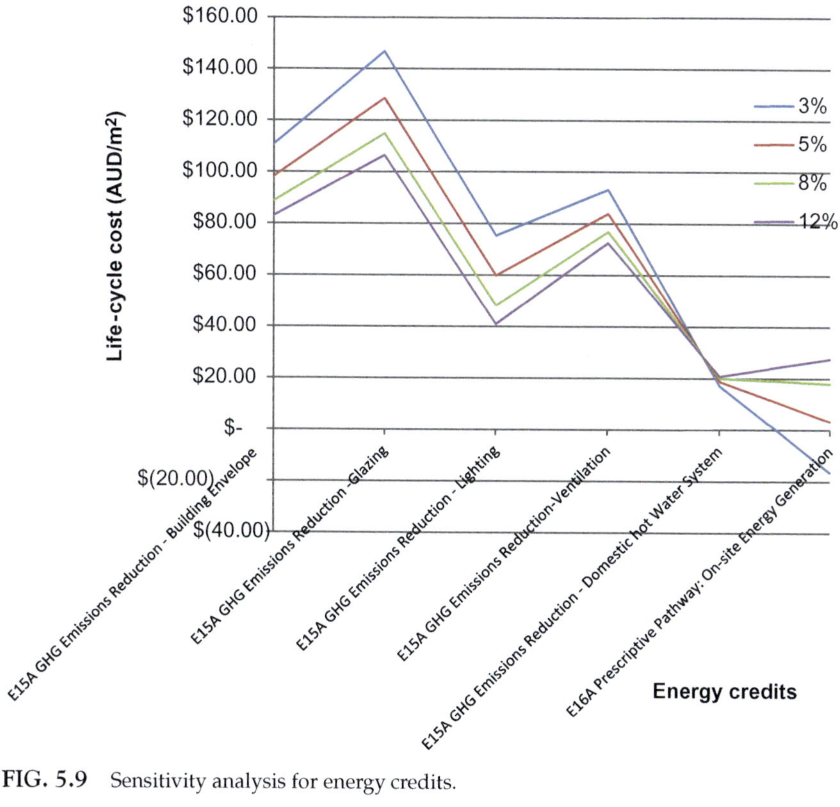

FIG. 5.9 Sensitivity analysis for energy credits.

5.7 Transport key criterion

As the name suggests, the main aim of this key criterion is to reduce the carbon emissions arising out of the occupant's travel, to and from the building. There is one prescriptive pathway available for this key criterion, and that allocated up to seven credit points.

All the credits in the transport key criterion only have an initial cost. Both "T17B.1 Access by Public Transport" and "T17B.5 Walkable neighborhood" credits heavily depend on the location of the building. If the building is located closer to a transport hub, and if it is closer to other amenities, both these credits can be achieved easily. However, a plot of land which is closer to public transport facilities and all other amenities will be expensive. There is a premium price to be paid while acquiring such a plot of land, and also, the price of land varies depending on the CBD, as well. This premium price is not considered in this life-cycle cost calculation.

While calculating costs for "T17B.2 Reduced Car Parking Provision" credit, a carpark is assumed to suit all the required criteria with a number of car parking slots to the given occupancy for an office building. However, there is only an initial cost included in the calculation, because the general maintenance is included and there is no specific maintenance available for a specific parking spot. Further, while deciding the demolition cost that is attributed to building elements, a separate cost for demolishing a car parking slot is not calculated. The calculation of "T17.B.4 Active Transport Facilities" is also similar to this. The "T17B.3 Low Emission Vehicle Infrastructure" fulfilled the option of providing 15% of parking for fuel efficient vehicles only.

As no costs contribute to the operational stage of the building life-cycle, there is no impact on the life-cycle cost from the changes in the discount rate. These life-cycle cost values are absolute, and only change with time. However, the lack of consideration of the premium land value is a limitation in this key criterion, and the user should be aware of this. Therefore, this is noted in the proposed life-cycle cost model. All the credits of this key criterion incur only initial costs. Therefore there are no specific life-cycle cost formulae given for this section. Table 5.10 reports the life-cycle cost ranges across main CBDs for discount rate 3.25%.

According to Table 5.10, there are many credits with lower life-cycle costs, except for "T17B.2 Reduced Car Parking Provisions" which has a comparatively higher cost.

5.7.1 Optimum solutions for transport credits

Fig. 5.10 below reports the optimum solutions for transport credits. The life-cycle cost is for Sydney considering a 3.25% discount rate. "T17B.1 Access by Public Transport" is the optimum solution in this key criterion,

TABLE 5.10 Life-cycle cost for transport credits.

Credit	Credit point	Life-cycle cost range across main CBDs (AUD/m^2)
T17B.1 Access by Public Transport	3	3.55–4.68
T17B.2 Reduced Car Parking Provisions	1	3.21–3.91
T17B.3 Low Emission Vehicle Infrastructure	1	16.91–19.66
T17B.4 Active Transport Facilities	1	5.44–6.65
T17B.5 Walkable Neighborhood	1	1.20–1.48

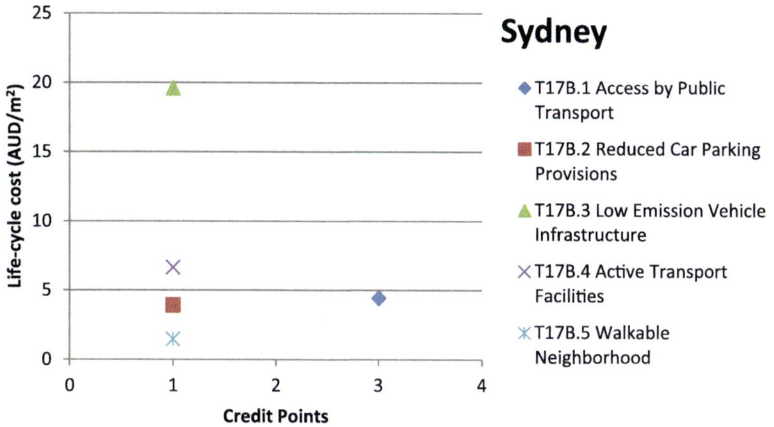

FIG. 5.10 Optimum solutions for transport credits.

with three credit points and lower life-cycle cost, as well. However, the following are the optimum selections for transport key criteria.

- "T17B.1 Access by Public Transport"—This credit requires a higher premium for land value. This is not included in the life-cycle cost
- "T17B.5 Walkable Neighborhood"
- "T17B.4 Active Transport Facilities"
- "T17B.2 Reduced Car Parking Provisions"

However, it necessary to identify that these credits significantly depend on the location of the plot of land.

Fig. 5.11 below reports the changes in life-cycle cost of transport credits in the CBDs. There are no significant changes to life-cycle costs across the CBDs. However, as mentioned in the previous section, this life-cycle cost calculation does not consider the premium price to be paid on cost of land due to the location. If this cost component is added to the life-cycle cost,

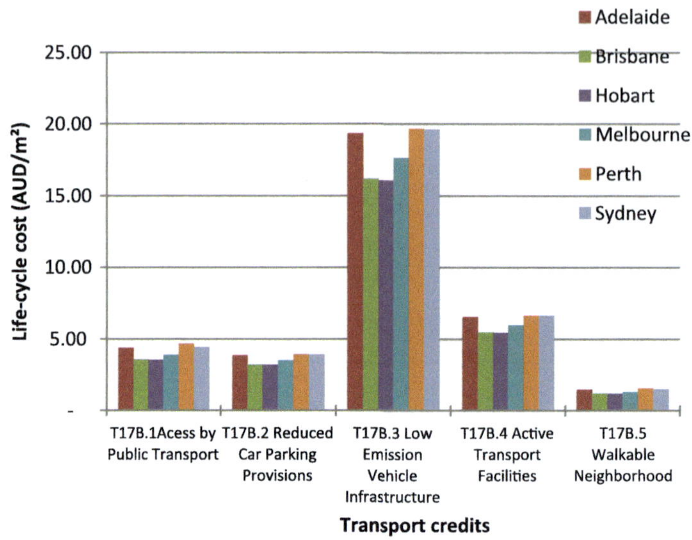

FIG. 5.11 Life-cycle cost for transport credits in six CBDs.

changes can be anticipated in the life-cycle cost, because the cost of land significantly changes across various CBDs in Australia.

5.8 Water key criterion

By using the prescriptive pathway, the proposed model can achieve up to six credit points through the water key criterion. The main aim of this key criterion is to minimize potable water consumption (Green Building Council of Australia, 2015). To achieve this aim, the users can use water efficient sanitary fixtures, reuse rainwater, avoid using water for heat rejection in HVAC systems, use water efficient landscape irrigation, and so on. Table 5.11 reports the cost contributions of each of the cost components to the life-cycle cost.

According to Table 5.11, all the credits make contributions to all the three types of life-cycle costs. For "W18B.1 Prescriptive Pathway: Sanitary Fixture Efficiency" credit calculations, all the fixtures are water efficient, and within the required rating levels. Further, while calculating the water costs within the life-cycle of the building, the water savings predicted from using water efficient sanitary fixtures are considered.

5.8.1 Life-cycle cost calculations for water credits

While calculating life-cycle costs for "W18B.2 Prescriptive Pathway: Rainwater Reuse," a 75 kL water tank is considered. Further, for "W18B.4 Prescriptive Pathway: Landscape Irrigation" a drip irrigation system is

TABLE 5.11 Cost components of credits in water key criterion.

Credit	Initial cost	Maintenance, replacement and other costs	Demolition cost
W18B.1 Prescriptive Pathway: Sanitary Fixture Efficiency	✓	✓	✓
W18B.2 Prescriptive Pathway: Rainwater Reuse	✓	✓	✓
W18B.3 Prescriptive Pathway: Heat Rejection	✓	✓	✓
W18B.4 Prescriptive Pathway: Landscape Irrigation	✓	✓	✓
W18B.5 Prescriptive Pathway: Fire System Test Water	✓	✓	✓

TABLE 5.12 Equations for water credits.

$\text{Life}-\text{cycle cost for sanitary fixtures} = \textit{Initial cost for sanitary fixtures including fixing} + \textit{Annual maintenance cost} \times \textit{PVA} - \textit{annual water savings from fixtures} \times \textit{PVA} + \textit{NPV of cost of demolition}$	Eq. (5.25): Life-cycle cost for sanitary fixtures
$\text{Life}-\text{cycle cost for rainwater tank} = \textit{Initial cost for rainwater tank} + (\textit{Nr of hours for annual cleaning} \times \textit{Nr of labourers} \times \textit{Hourly labour rate}) \times \textit{PVA} - \textit{Annual water savings} + \textit{NPV of cost of demolition}$	Eq. (5.26): Life-cycle cost for rainwater tank
$\text{Life}-\text{cycle cost for drip irrigation system} = \textit{Initial cost of drip irrigation system} + (\textit{Annual maintenance cost} \times \textit{PVA}) + \textit{NPV of cost of demolition}$	Eq. (5.27): Life-cycle cost for domestic hot water system

considered. Further, Table 5.12 provides the formulae used for calculating the life-cycle cost for water credits.

All the sanitary fixtures follow the required water efficiency labelling system (WELS). Therefore, the water savings are based on the saving rates given by WELS. After considering these formulae given in Table 5.12, the life-cycle cost is calculated and reported in Table 5.13.

According to Table 5.13, depending on the type of ventilation system in the building, there are two options available for "W18B.3 Heat Rejection." However, once again, using natural ventilation is considered to have lesser life-cycle cost.

TABLE 5.13 Life-cycle cost for water credits.

Credits	Credit point	Life-cycle cost (AUD/m^2)					
		Adelaide	**Brisbane**	**Hobart**	**Melbourne**	**Perth**	**Sydney**
W18B.1 Sanitary Fixture Efficiency	1	84.96	71.06	70.29	77.24	86.04	85.97
W18B.2 Rainwater Reuse	1	2.90	2.42	2.40	2.63	2.94	2.93
W18B.3 Heat Rejection—Naturally Ventilated	2	2.90	1.46	1.16	4.90	9.69	9.63
W18B.3 Heat Rejection—Mechanically Ventilated—Air Cooled System	2	147.38	144.96	145.93	141.09	152.70	146.41
W18B.4 Landscape Irrigation	1	0.62	0.52	0.51	0.56	0.63	0.63
W18B.5 Fire System Test Water	1	42.10	41.60	41.50	41.50	50.90	42.70

5.8.2 Optimum solutions for water credits

There are a total of five credits considered in the water key criterion, including two options for heat rejection. However, the optimum solutions are the credits with the lowest life-cycle cost and highest number of credit points. Fig. 5.12 presents the optimum solution for water credits calculated for Sydney with the discount rate being 3.25%.

According to Fig. 5.12, the optimum credits in the water key criterion are follows:

- "W18B.3 Heat Rejection—Naturally Ventilated"
- "W18B.4 Landscape Irrigation"—Drip irrigation is considered
- "W18B.2 Rainwater Reuse"
- "W18B.1 Sanitary Fixture Efficiency"

When there is a natural ventilation system in the building, "W18B.3 Heat Rejection" can be directly achieved (refer Section 4.4).

Apart from the optimum solutions for Sydney, Fig. 5.13 illustrates the life-cycle cost for water credits in different CBDs.

According to Fig. 5.13, there are no major discrepancies across the CBDs in Australia for water credits.

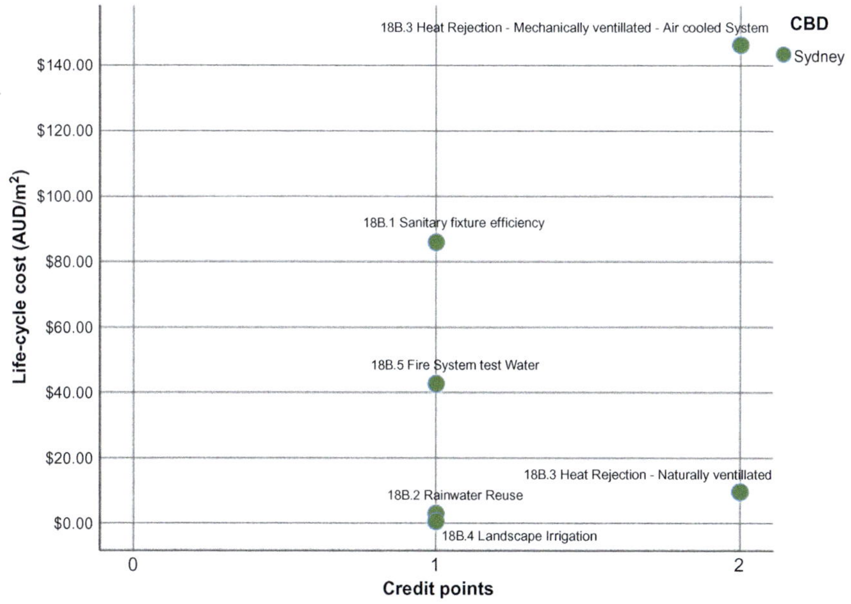

FIG. 5.12 Optimum solutions for water credits.

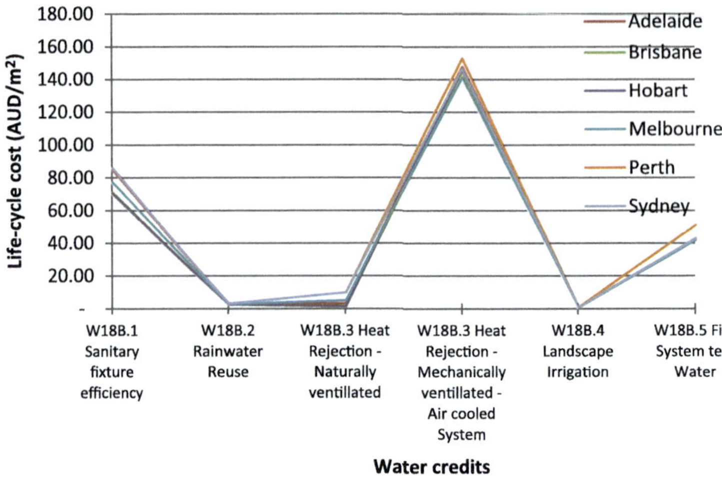

FIG. 5.13 Life-cycle cost for water credits in six CBDs.

5.8.3 Sensitivity analysis for water credits

The credits in the water key criterion have an impact on all the different components of life-cycle costs. Therefore, it is worth identifying the sensitivity of these costs to the discount rate. Fig. 5.14 below reports the sensitivity of the life-cycle costs to the changes in the discount rate.

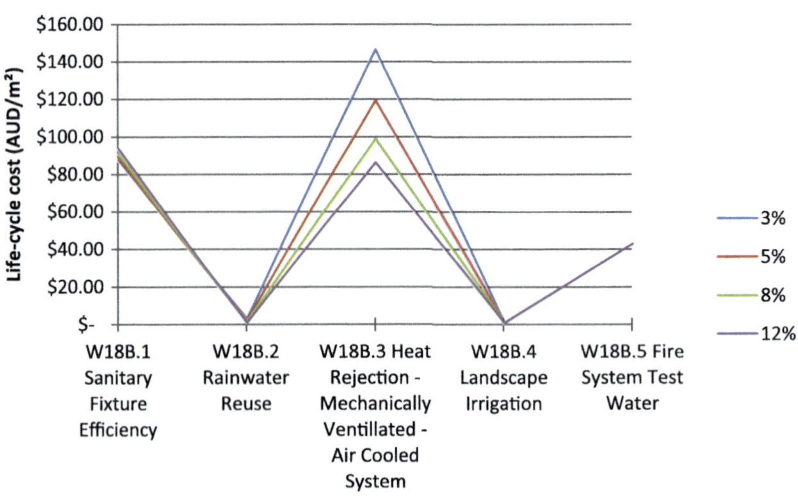

FIG. 5.14 Sensitivity analysis for water credits.

In Fig. 5.14, to achieve "W18B.3 Prescriptive Pathway: Heat Rejection," a HVAC system using air for heat rejection is used. This system incurs regular maintenance, replacement, and demolition costs. Therefore, there are lots of costs depending on the discount rate. This leads to a drastic change in life-cycle cost as a result of the changes in discount rate. "W18B.2 Prescriptive Pathway: Rainwater Reuse" and "W18B.4 Prescriptive Pathway: Landscape Irrigation" credits can be achieved using a minimum life-cycle cost, when compared to the other credits in the water key criterion.

The sensitivity to the discount rate is significantly displayed in Fig. 5.14. As an example, if the discount rate is 3%, out of "W18B.1 Prescriptive Pathway: Sanitary Fixture Efficiency" and "W18B.3 Prescriptive Pathway: Heat Rejection," using water efficient sanitary fixtures credit is clearly cheaper, as per the life-cycle cost (refer Fig. 5.14). However, if the discount rate is 12%, "W18B.3 Prescriptive Pathway: Heat Rejection" is slightly cheaper than it was in the previous scenario. Therefore, this scenario illustrates the importance of using the best suited discount rate for life-cycle cost calculations.

5.9 Material key criterion

As the name suggests, this key criterion focuses on building materials. According to the Green Building Council of Australia (2015), the aim of this key criterion is to address the consumption of resources in a building construction context, by encouraging the selection of lower impact material. The main focus of this key criterion is on material, namely, concrete, steel, timber, and the minimum use of PVC. Table 5.14 reports the contribution of costs towards the life-cycle cost.

According to Mori and Ellingwood (1993), the long-term strength of concrete changes due to maturity. Further environmental stressors may attack the integrity of concrete and/or steel reinforcement in concrete with or independent of operating, environmental, and accidental loads (Mori & Ellingwood, 1993) causing the strength to degrade over time. Therefore, according to Illankoon, Tam, Le, and Wang (2018), it can be argued that concrete only needs to be maintained due to certain external issues, such as accidental loads and severe environmental conditions. Further, usually in a building, crack formation is mostly due to design failures or technical failures in placing and curing concrete, and other than that, in normal conditions, specific maintenance is not required for concrete (Illankoon, Tam, Le, et al., 2018). Therefore, while using SCM in concrete for obtaining "Mat19B Prescriptive Pathway—Life cycle Impacts" maintenance and replacement costs are not included. A similar argument can be used for "Mat 20.1 Structural Reinforcing Steel" as well.

TABLE 5.14 Cost components of credits in material key criterion.

Credit	Initial cost	Maintenance, replacement and other costs	Demolition cost
Mat19B Prescriptive Pathway—Life cycle Impacts	✓	–	✓
Mat 20.1 Structural Reinforcing Steel	✓	–	✓
Mat 20.2 Timber Products	✓	✓	✓
Mat 20.3 Permanent, Formwork, Pipes, Flooring, Blinds and Cables	✓	✓	✓
Mat 21 Product Transparency and Sustainability	✓	–	–
Mat 22 Reduction of Construction and Demolition Waste	✓	–	–

Since the building is a concrete framed building, for this credit, only reinforcement steel in concrete is considered. Therefore, there will only be an initial cost and a demolition cost, except in extreme conditions, where the structure needs major renovation. As a result, "Mat 20.1 Structural Reinforcing Steel" credit does not include any maintenance, replacement, and other costs.

There are many SCM materials that can be used. However, according to Illankoon, Tam, Le, et al. (2018) 50% of cement replaced with slag is the optimum solution for SCM replacement. Therefore, this option is used for life-cycle calculation in the proposed model. Further, the regular maintenance costs included in "Mat 20.2 Timber Products" and in "Mat 20.3 Permanent, Formwork, Pipes, Flooring, Blinds and Cables" are at a minimum. As a result of this, the maintenance, replacement, and other costs are minimized in this key criterion. Therefore, the impact on life-cycle cost due to the changes in discount rate is at a minimum. Further, there are no specific formulae used for life-cycle cost calculations in this key criterion. Table 5.15 reports the life-cycle cost for material credits at a 3.25% discount rate.

Various credits are discussed under the materials key criterion. The life-cycle cost and the credit points vary significantly in this key criterion. "Mat19B.1 Life Cycle Impacts—Concrete" has the highest life-cycle cost, but this credit also achieves three credit points. Further, "Mat22A Reduction of Construction and Demolition Waste" creates significant change in the life-cycle cost depending on the CBD (refer Table 5.15).

TABLE 5.15 Life-cycle cost for material credits.

Credits	Credit point	Life-cycle cost (AUD/m²)					
		Adelaide	Brisbane	Hobart	Melbourne	Perth	Sydney
Mat19B.1 Life-Cycle Impacts—Concrete	3	246.10	205.83	201.36	223.73	219.21	214.01
Mat 19B.2B Reduced Use of Steel Reinforcement	2	37.85	40.70	37.31	41.00	43.10	51.05
Mat 20.1 Structural reinforcing Steel	1	1.46	1.20	1.18	1.30	1.56	1.48
Mat 20.2 Timber Products	1	21.30	17.81	17.62	19.36	21.57	21.55
Mat 20.3 Permanent, Formwork, Pipes, Flooring, Blinds and Cables	1	1.46	1.20	1.18	1.30	1.56	1.48
Mat 21 Product Transparency and Sustainability	3	4.39	3.59	3.55	3.90	4.68	4.44
Mat 22A Reduction of Construction and Demolition Waste	1	27.29	27.31	14.21	15.61	30.87	– 9.95

5.9.1 Optimum solutions for material credits

According to Table 5.15, there are credits with higher life-cycle costs and higher credit points. Therefore, considering these factors, Fig. 5.15 reports the optimum solutions in material credits, for Sydney at a 3.25% discount rate.

According to Fig. 5.15, the optimum solutions are as follows:

- "Mat21 Product Transparency and Sustainability"—this credit is not a standalone credit. This credit can only be achieved while certain other credits are fulfilled. These inter-dependencies are considered in the proposed model
- "Mat22A Reduction of Construction and Demolition Waste"—this credit depends significantly on the CBD
- "Mat20.3 Permanent, Formwork, Pipes, Flooring, Blinds and Cables"
- "Mat20.2 Timber Products"
- "Mat20.1 Structural Reinforcing Steel"

As mentioned above, construction demolition waste credit depends on the CBD. Therefore, it is necessary analyze the life-cycle cost in various CBDs as well.

Fig. 5.16 illustrates the life-cycle cost of material credits across various CBDs. According to Fig. 5.16, there are considerable changes in the

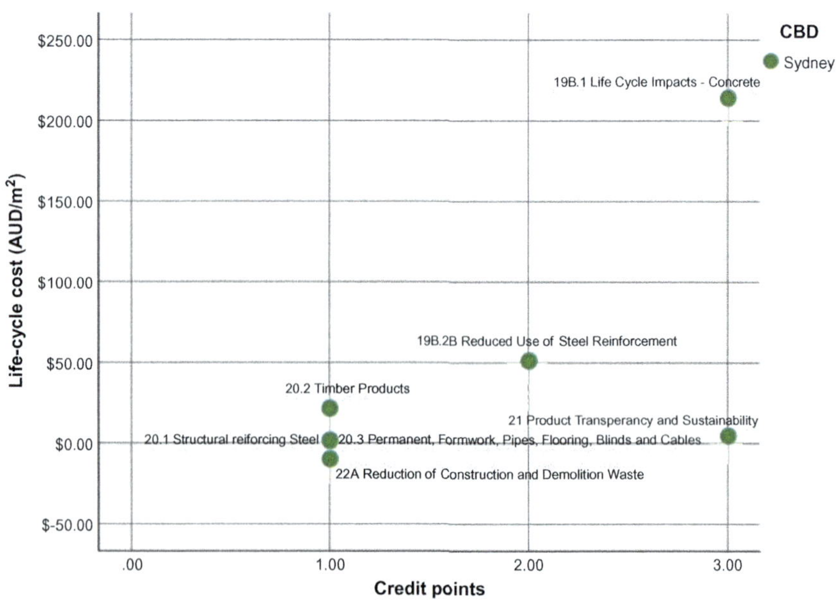

FIG. 5.15 Optimum solutions for material credits.

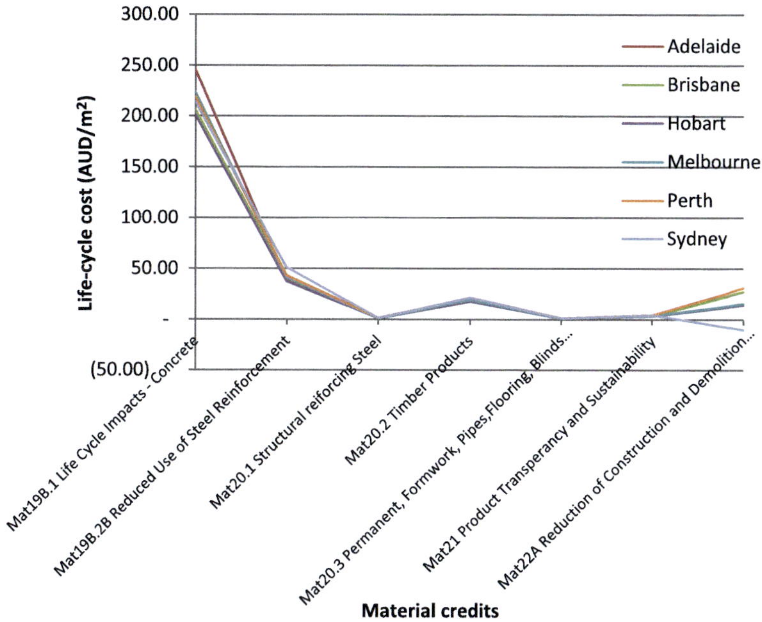

FIG. 5.16 Life-cycle cost for material credits in six CBDs.

life-cycle cost of construction waste credits. The main reason is the cost of disposing the construction and demolition waste. In Sydney, waste disposal is expensive. Therefore, when the construction and demolition waste is reduced, it is illustrated as a saving (refer Fig. 5.16).

5.10 Land use and ecology key criterion

According to the Green Building Council of Australia (2015), the aim of this key criterion is to reduce the negative impacts on the ecological value of sites as a result of urban development, and to reward projects that minimize harm and enhance the quality of the local ecology. Therefore, the credits of this key criterion look into the state of the plot of land that the building is constructed upon. Most of the credits in this key criterion focus on the initial cost and few credits have an impact on initial cost, maintenance cost and the demolition costs. Table 5.16 reports the contribution of each of credit to the life-cycle cost.

All the credits, except for "L23.1 Ecological Value" and "L25 Heat Island Effect Reduction" have only initial costs. "L23.1 Ecological Value" is achieved when the ecological value of the site is improved through vegetation, using water bodies, and so on. However, for the

TABLE 5.16 Cost components of credits in land use and ecology key criterion.

Credit	Only initial cost is required	All three cost components are required[a]
L23.0 Endangered, Threatened or Vulnerable species	✓	–
L23.1 Ecological Value	–	✓
L24.0 Conditional Requirement	✓	–
L24.1 Reuse of Land	✓	–
L24.2 Contamination and hazardous Material	✓	–
L25 Heat Island Effect Reduction	–	✓

[a] *All three components of cost include initial cost, maintenance, replacement and other costs and demolition costs.*

purpose of the life-cycle cost calculation, the ecological value is improved by having vegetation and constructing an artificial water body within the site. Both these included regular maintenance throughout the life-cycle and further, these incurred demolition costs, as well. Similarly, "L25 Heat Island Effect Reduction" is achieved by reducing the impact of the heat island effect, through vegetation, green roofs, hot water and PV panels, and so on. For the purpose of this calculation, authors used vegetation, PV panels, and water bodies to reduce the impact of the heat island.

The condition of the land at the time of purchase significantly influence credits, such as "L23.0 Endangered, Threatened or Vulnerable Species," "L24.0 Conditional Requirement," "L24.1 Reuse of Land," and "L24.2 Contamination and Hazardous Material." As an example, if there is no contamination in the land at the time of purchase, "L24.2 Contamination and Hazardous Material" cannot be achieved. Further, if the land cannot satisfy the conditional requirements such as "L23.0 Endangered, Threatened or Vulnerable species" and "L24.0 Conditional Requirement," most of the underlying credits cannot be awarded, either. Therefore, the users must be extremely careful in the selection of the land during the initial decision-making stage. Table 5.17 reports the life-cycle cost for land use and ecology credits at a discount rate of 3.25%.

According to Table 5.17, credits such as "L23.0 Endangered, Threatened or Vulnerable Species," "L24.0 Conditional Requirement," and "L24.1 Reuse of Land" all have the same values. The main reason is that the value of the land is not included in the calculation, and only the documentation cost is included.

TABLE 5.17 Life-cycle cost for land use and ecology credits.

Credit	Credit point	Life-cycle cost range across main CBDs (AUD/m^2)
L23.0 Endangered, Threatened or Vulnerable Species	Required	1.20–1.56
23.1 Ecological Value	3	9.20–11.26
24.0 Conditional Requirement	Required	1.20–1.56
24.1 Reuse of Land	1	1.20–1.56
24.2 Contamination and Hazardous Material	1	66.07–80.61
25 Heat Island Effect Reduction	1	7.10–8.69

5.10.1 Optimum solutions for land use and ecology credits

Fig. 5.17 presents the optimum solutions for land use and ecology credits for Sydney at a discount rate of 3.25%. Based on Fig. 5.17, the optimum solutions are as follows:

- "L23.1 Ecological Value"—this credit was a conditional credit

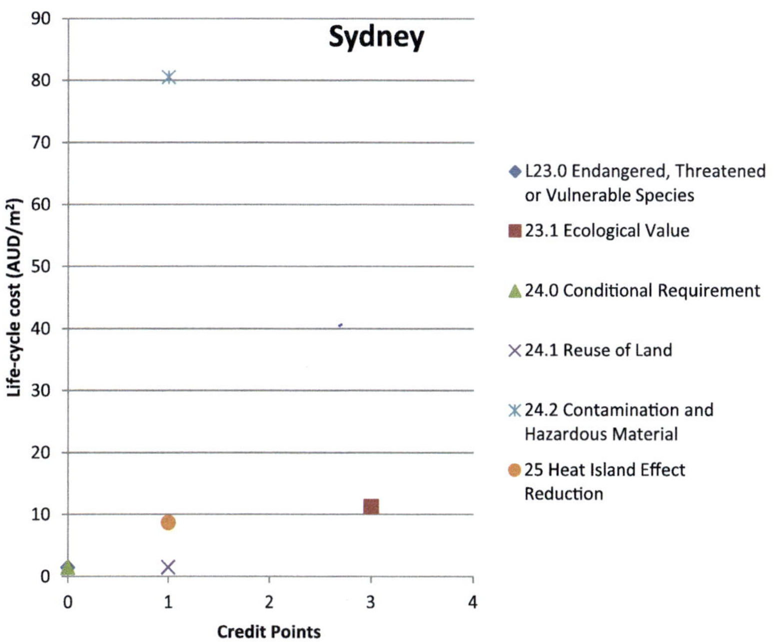

FIG. 5.17 Optimum solutions for land use and ecology credits.

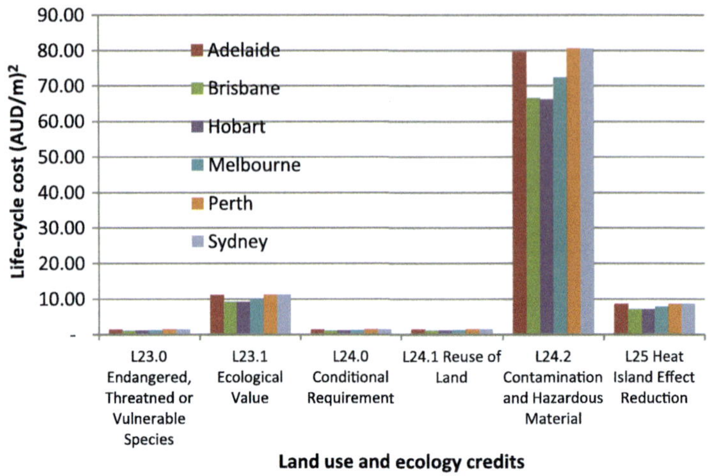

FIG. 5.18 Life-cycle cost for land use and ecology credits in six CBDs.

- "L24.1 Reuse of Land"—the land value is not included in the life-cycle cost
- "L25 Heat Island Effect Reduction"

The life-cycle cost and optimum solutions significantly depend on the land value and the state of land at the time of purchase.

Fig. 5.18 reports the changes in life-cycle cost of land use and ecology credits across the CBDs. According to Fig. 5.18, there are no considerable changes in the CBDs. However, if the land value is included in the cost calculations, there can be possible changes in life-cycle cost across the CBDs, because, the land cost significantly changes across the CBDs.

5.11 Emissions key criterion

This key criterion mainly focuses on the pollution targets of the building. The main aim of this key criterion is to assess the environmental impacts of point source pollution generated by projects (Green Building Council of Australia, 2015). Therefore, this key criterion uses many credits to prevent pollution at the point of emergence. Table 5.18 reports the different cost components included in the life-cycle cost calculation.

All the credits in this key criterion make a contribution to maintenance, demolition, replacement, and other costs (refer Table 5.18). To achieve the "Em26.1 Reduced Peak Discharge" and "Em26.2 Reduced Pollution Targets" an efficient storm water management system is considered. Further, "Em27.0 Light Pollution to Neighboring Bodies" and "Em27.1 Light Pollution to Night Sky" directly relate to the provision of external lighting.

TABLE 5.18 Cost components of credits in emissions key criterion.

Credit	Initial cost	Maintenance, replacement and other costs	Demolition cost
Em26.1 Reduced Peak Discharge	✓	✓	✓
Em26.2 Reduced Pollution Targets	✓	✓	✓
Em27.0 Light Pollution to Neighbouring Bodies	✓	✓	✓
Em27.1 Light Pollution to Night Sky	✓	✓	✓
Em28 Legionella Impacts from Cooling Systems	✓	✓	✓
Em29 Refrigerant Impact	✓	✓	✓

5.11.1 Life-cycle cost for emissions credits

There are various formulae used in calculating the life-cycle cost. Table 5.19 reports the specific formulae used in calculating life-cycle cost for emissions credits. Life-cycle cost is calculated and reported in Table 5.20, for Sydney, at a discount rate of 3.25%.

According to Table 5.20, the life-cycle cost ranges across a wide spectrum of emission credits. However, for credits such as "Em28 Legionella Impacts from Cooling Systems" and "Em29 Refrigerant Impact," there are options available, depending on the type of ventilation in the building.

5.11.2 Optimum solutions for emission credits

Fig. 5.19 reports the optimum solutions for emissions credits. The life-cycle costs are calculated for Sydney and the discount rate is 3.25%. Based on Fig. 5.19, the optimum solutions for emission criterion are as follows:

- "Em28 Legionella Impacts from Cooling Systems—Naturally Ventilated/Air Cooled System"—if the ventilation system does not use water for heat rejection, this credit point is achieved directly. The cost included is the cost of documentation
- "Em29 Refrigerant Impact—Naturally Ventilated"—if the ventilation system does not use refrigerants, this credit point is achieved directly
- "Em27.1 Light Pollution to Night Sky"
- "Em26.1 Reduced Peak Discharge"
- "Em26.2 Reduced Pollution Targets"

TABLE 5.19 Equations for emissions credits.

Life – cycle cost for stormwater management system = *Initial cost for storm water management system* + *Annual maintenance cost* × *PVA* + *NPV of cost of demolition*	Eq. (5.28): Life-cycle cost for stormwater management system
Life – cycle cost for external lighting = *Initial cost for external light fittings, fixtures and lighting points* + (*Annual maintenance* × *PVA*) + *Replacement cost of lamps every* 3 *year* + *NPV of cost of demolition*	Eq. (5.29): Life-cycle cost for external lighting
Annual maintenance cost = [(*Nr of hours for annual maintenance* × *Nr of labourers* × *Hourly labour rate*) + (*Nr of hours for annual maintenance* × *Nr of electricians* × *Hourly rate for Electrician*)] + (*Nr of hours for monthly visual inspection and cleaning for lighting fixtures and controls* × *Hourly labour rate* × 12)	Eq. (5.30): Annual maintenance for external lighting

Similar to previous optimum solutions in the IEQ credits and water credits, in this section also, natural ventilation system is selected as the optimum solution.

The optimum solutions are selected based on the life-cycle costs calculated in Sydney. Therefore, Fig. 5.20 reports the life-cycle costs across other CBDs.

According to Fig. 5.20, there are no considerable changes in life-cycle costs across different CBDs.

5.11.3 Sensitivity analysis for emissions credits

All the emission credits contribute to maintenance costs associated with the system, and therefore, there is an impact on the life-cycle cost calculation, with changes in the discount rate. The analysis of sensitivity to the discount rate is given in Fig. 5.21 below.

According to Fig. 5.21, there are slight changes in the life-cycle cost as a result of the changes in the discount rate. The main reason for this is the minimal contribution of the costs incurred at the operational stage of the building. Thus, for this key criterion, although all the credits occur, the impact of the changes in discount rate on the life-cycle cost is insignificant.

5.12 Summary

This chapter discussed the life-cycle cost calculations for each credit in each key criterion. According to Section 5.2, there are three components of the life-cycle cost, namely, initial cost, maintenance, replacement and other

TABLE 5.20 Life-cycle cost calculations for emissions credits.

Credits	Credit point	Life-cycle cost (AUD/m^2)					
		Adelaide	Brisbane	Hobart	Melbourne	Perth	Sydney
Em26.1 Reduced Peak Discharge	1	53.29	44.57	44.08	48.44	53.97	53.92
Em26.2 Reduced Pollution Targets	1	79.93	66.85	66.12	72.66	80.95	80.88
Em27.0 Light Pollution to Neighbouring Bodies	Required	14.70	13.98	14.27	12.82	16.30	14.41
Em27.1 Light Pollution to Night Sky	1	18.91	17.97	18.34	16.48	20.96	18.53
Em28 Legionella Impacts from Cooling Systems—Naturally Ventilated	1	1.46	1.20	1.18	1.30	1.56	1.48
Em28 Legionella Impacts from Cooling Systems—Mechanically—Water Cooled system Ventilated	1	62.11	62.19	62.16	62.33	61.92	62.14
Em28 Legionella Impacts from Cooling Systems—Mechanically—Air Cooled system Ventilated	1	1.46	1.20	1.18	1.30	1.56	1.48
Em29 Refrigerant Impact—Naturally Ventilated	1	1.46	1.20	1.18	1.30	1.56	1.48
Em29 Refrigerant Impact—Mechanically Ventilated—Water Cooled system	1	217.37	217.67	217.55	218.16	216.71	217.49
Em29 Refrigerant Impact—Mechanically Ventilated—Air Cooled system	1	294.75	289.92	291.85	282.18	305.39	292.82

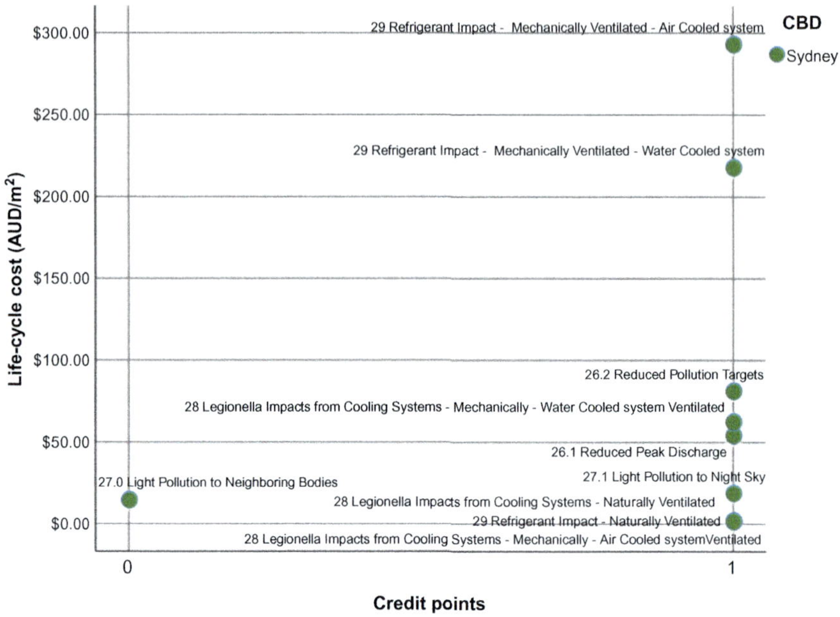

FIG. 5.19 Optimum solutions for emissions credits.

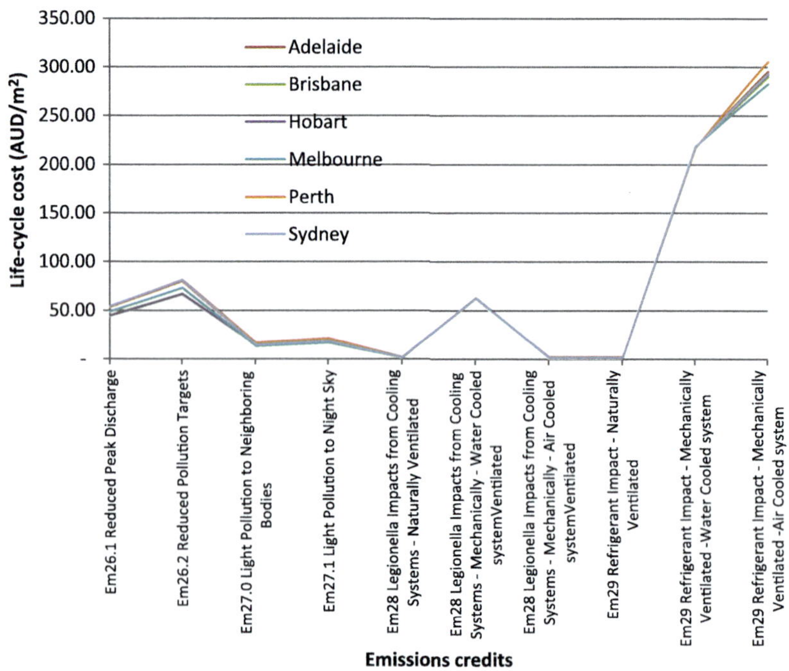

FIG. 5.20 Life-cycle cost for emissions credits in six CBDs.

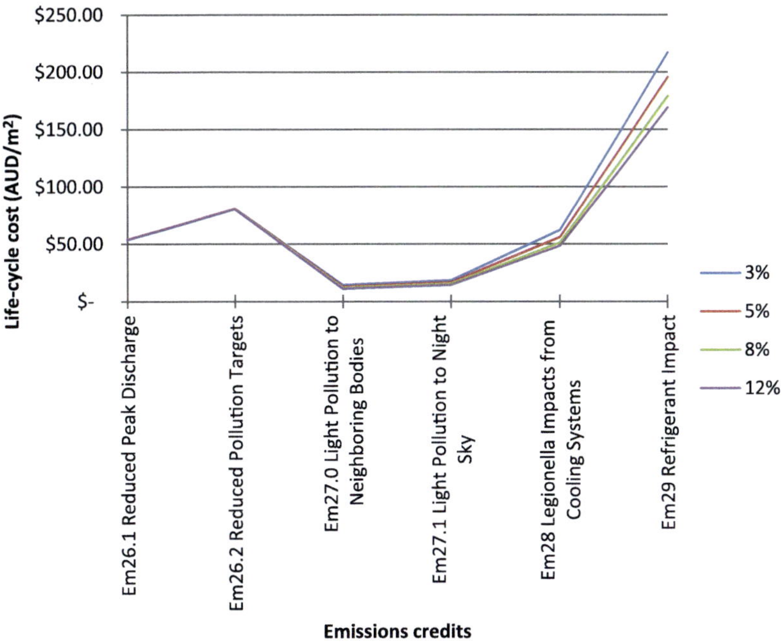

FIG. 5.21 Sensitivity analysis for emissions credits.

costs, and demolition costs. These three cost components are considered in calculating the life-cycle cost formula (refer Eq. 5.3). After considering all these life-cycle costs, the proposed model multiplies it by the relevant regional index, to arrive at the life-cycle cost for a specific regional area.

However, while considering different criteria, specific formulae are developed to calculate the life-cycle cost. All these formulae are illustrated in this chapter. Based on these formulae, the model calculates the life-cycle cost. However, this chapter provides an example of life-cycle cost for each credit, for Sydney at a discount rate of 3.25%. Following this, the optimum solutions are discussed for each key criterion. Further, the life-cycle costs of the CBDs are also analyzed.

Finally, a sensitivity analysis is carried out considering the changes in the discount rate. There are certain credits that make contributions to all the three cost components. All the key criteria with such types of credits were identified to perform the sensitivity analysis. Since sensible assumptions were made to develop the life-cycle costs, these assumptions are also mentioned in this chapter.

References

Australian Government. (2016). *Australian Government Bonds*. Retrieved from http://australiangovernmentbonds.gov.au/etbs/list-of-etbs/.

Australian National Audit Office. (2001). *Life cycle costing—Better practice guide*. Retrieved from http://www.anao.gov.au/uploads/documents/Life_Cycle_Costing.pdf.

Building Construction Authority. (2013). *BCA Green Mark for new non-residential buildings version NRB/4.1*. Retrieved from https://www1.bca.gov.sg/.

Building Research Establishment Environment Assessment Method. (2014). *BREEAM international new construction technical manual (SD 5075 1.0:2013)*. Retrieved from https://www.breeam.com/.

Dell'Isola, A. J., & Kirk, S. J. (2003). *Life cycle costing for facilities: economic analysis for owners and professionals in planning, programming, and real estate development: designing, specifying, and construction, maintenance, operations, and procurement*. Kingston, MA: Reed Construction Data.

Deloitte Access Economics. (2016). *Construction sector—Outlook, labour costs and productivity*. Retrieved from: http://www.mbav.com.au/vpLink.aspx?ID=3&D=000000007858&G=0fEKM8zNyUKqaljeGEtfcA2&N=YES.

Department of Climate Change and Energy Efficiency. (2012). *Baseline energy consumption and greenhouse gas Emissions in commercial buildings in Australia—Part 1*. Department of Climate Change and Energy Efficiency Retrieved from https://www.energy.gov.au/publications/baseline-energy-consumption-and-greenhouse-gas-emissions-commercial-buildings-australia.

Green Building Council of Australia. (2015). *Green Star—Design and As Built version 1.1*. Retrieved from: https://www.gbca.org.au/green-star/green-star-design-as-built/the-rating-tool/.

Green Building Council of Australia. (2018). *About GBCA*. Retrieved from: http://www.gbca.org.au/about/.

Illankoon, I. M. C. S., Tam, V. W., & Le, K. N. (2018). Life cycle costing for insulated pitched roof structures. In *Proceedings of the 21st International Symposium on Advancement of Construction Management and Real Estate* (pp. 981–993). Springer.

Illankoon, I. M. C. S., Tam, V. W. Y., Le, K. N., & Shen, L. (2016). Cost premium and the life cycle cost of green building implementation in obtaining green star rating in Australia. In *Paper Presented at the Pacific Association of Quantity Surveyors Congress (PAQS), Christchurch, New Zealand*.

Illankoon, I. M. C. S., Tam, V. W. Y., Le, K. N., & Wang, J. Y. (2018). Life cycle costing for obtaining concrete credits in green star rating system in Australia. *Journal of Cleaner Production, 172*, 4212–4219. https://doi.org/10.1016/j.jclepro.2017.11.202.

International Organisation for Standardization [ISO]. (2017). *Buildings and constructed assets—Service life planning Part 5: Life cycle costing (Vol. ISO 15686-5)*. Switzerland: International Standard Organization [ISO].

Mori, Y., & Ellingwood, B. R. (1993). Reliability-based service-life assessment of aging concrete structures. *Journal of Structural Engineering (United States), 119*(5), 1600–1621. https://doi.org/10.1061/(ASCE)0733-9445(1993)119:5(1600).

Rawlinsons. (2017). *Australian construction Handbook. Vol. 34*. Perth, Western Australia: Rawlinson Publishing.

Stanford, H. W. (2010). *Effective building maintenance: protection of capital assets*. Lilburn, GA/Boca Raton, FL: Fairmont Press.

Tam, V. W. Y., Le, K. N., Zeng, S. X., Wang, X., & Illankoon, I. M. C. S. (2017). Regenerative practice of using photovoltaic solar systems for residential dwellings: An empirical study in Australia. *Renewable and Sustainable Energy Reviews, 75*, 1–10. https://doi.org/10.1016/j.rser.2016.10.040.

United States Green Building Council. (2014). *LEED V4 for building design and construction*. Washington, DC: US Green Building Council.

Wong, I. L., Perera, S., & Eames, P. C. (2010). Goal directed life cycle costing as a method to evaluate the economic feasibility of office buildings with conventional and TI-façades. *Construction Management and Economics, 28*(7), 715–735. https://doi.org/10.1080/01446191003753867.

CHAPTER

6

Life-cycle cost model development

6.1 Introduction

After developing the life-cycle costs and specifying the inter-dependencies among credits, this study focused on developing the proposed life-cycle cost model for optimum selection of credits. The model is developed based on Java software. The user has to input data and specify the constraints. Once the user inputs are given, the proposed model selects optimum solutions to obtain desired ratings. However, this model only includes 74 credit points after considering 63 credits in all (refer Chapter 4 for selection and elimination of credits). Therefore, this proposed model only works up to 5-star rating level.

This model is developed based on a series of user constraints and inter-dependencies of credits in Green Star Design and As-Built version 1.1. The model selects the optimum solutions by minimizing the life-cycle costs while selecting the credits with higher credit points. This chapter provides an in-depth illustration of the development of the model.

6.2 Introduction to the life-cycle cost model

The life-cycle cost model selects optimized solutions to achieve the desired Green Star ratings. The main parameters of the model are the life-cycle cost and the value of credit points. The main aim of this model is to minimize the life-cycle cost while achieving the highest number of credit points. The programming language used to develop this model is Java. Java is an object-oriented programming language (Naressi et al., 2001). Further, Java is claimed to be platform-independent allowing the software to run in (Naressi et al., 2001). According to Currie (2006), Java has a built in graphical user interface and the syntax is similar to C++ programming language, yet simpler in many ways. This book focuses on developing

https://doi.org/10.1016/B978-0-12-820062-9.00006-2

a life-cycle cost model to obtain optimum solution based on set requirements. Therefore, the authors had to develop the set of rules based on which the solutions are drawn. Afterwards, these rules are included to the selected program to arrive at the solutions. Author could effectively and efficiently set up source code file developed based on the set rules which are explained in detail in the latter sections, using the java software. Further, object-oriented nature of the programming language has act as an enabler to effectively set up the variables and develop the source code. Therefore, Java application is used effectively to develop the proposed life-cycle cost model.

This model can be used by the user in the initial decision making stages to identify the best credit combination optimizing the life-cycle cost of the green building. This model comprises various constraints of users, and the constraints and requirements of the Green Star rating tool. Once the model is used, it evaluates all the possible solutions. According to Section 4.3, this book considered 63 credits for the life-cycle cost model, with a total credit point value of 74. If the user requires a minimum of 45 credit points for certification, there are around 3.12×10^{20} (i.e., $^{74}C_{45}$) number of possible combinations as solutions. This shows the complexity involved in choosing the best possible solution manually. There is a very high possibility of choosing the solution which is not the optimum solution considering the life-cycle cost. Therefore, this life-cycle cost model processes millions of solutions, eliminates the options that do not cater to user constraints, combines and selects the credits with inter-dependencies, and finds the optimum solution with the lowest life-cycle cost, within the required level of credit points for certification.

The model is developed considering certain conditions, user constraints, inclusions and exclusions, and dependencies. Each of these are carefully considered and embedded in the life-cycle cost model. Based on that, the optimum solutions are selected. Therefore, the model development can be illustrated in five stages, namely, user inputs, inclusions and exclusions based on user inputs, dependencies among credits, conditions of selection, and the selection process. The following sections of this chapter discuss all these stages in model development.

6.3 User inputs to the life-cycle cost model

User inputs are of significant importance in this model. Users have various constraints, and solutions should be selected to satisfy these constraints. There are two types of user information that are required for this model. Certain user inputs comprise information meant to process the calculation, and others are user constraints. Therefore, initially, the input screen collects user information and then begins processing.

6.3.1 Identification of user information required for the model

There are four main types of information inputs included in the model. Initially, the user specifies the expected rating for the building. The model supports a four-star silver rating and a five-star gold rating. The four-star rating has a credit point range of 45–59 credit points, and the five-star rating ranges from 60 to 74 credit points. The credit point requirement is given as a range, and therefore, there can be many solutions that satisfy a given range of credit points. Therefore, the model provides three options to satisfy the lowest boundary of the range, the highest boundary and the middle of the range.

As illustrated in Chapter 5, this study developed life-cycle cost calculations for the six main CBDs in Australia. The user can choose the closest CBD to the project to obtain a more consistent and realistic output. Further, if the project is located in a regional area, the user can select the regional area from a dropdown list. If the user selects a regional area, the model adjusts the life-cycle cost data by using the regional indices published by the Rawlinsons (2016) database.

Discount rate is one of the significant parameters in this book. The life-cycle cost calculations are developed based on a discount rate of 3.25% (refer Section 5.2). The user can change the discount rate to suit the risk and other considerations, across a range, from 0.25% to 20%. Based on the discount rate provided by the user, the model calculates the life-cycle cost using the NPV technique (refer Section 5.2).

The expected Green star certification level, the location of the proposed green buildings and the discount rate are the user information required by the proposed life-cycle cost model. The proposed life-cycle cost model requires this information to re-calculate the life-cycle cost and to identify the level of certification.

6.3.2 Identification of user inputs/constraints required for the model

There are various constraints and decisions on part of the user that need to be considered in the life-cycle cost model. Therefore, in the user input screen, the user is required to answer a series of questions to identify their exact requirements and constraints. Based on the answers received, credits are included or eliminated from during the initial stage. A list of user information, inputs and constraints are given in Table 6.1.

All these questions related to user inputs and constraints are developed based on the requirements of the Green Star Design and As-Built version 1.1 rating tool. According to Table 6.1, the first four questions seek user information and questions 5 to 19 identify the user's constraints. There are questions related to the location of the building. In question 5, the model

TABLE 6.1 User inputs.

Questionnaire to the user	
Q1	Select the desired level of green building certification using Green Star Design and As-Built v1.1
Q2	Select the CBD in which the building is located
Q3	If the building is located in a regional area, please specify the region (If the building is in the CBD, select the CBD from the list)
Q4	Provide the discount rate (range: 0.25–20%)
Q5	Can the building users access the office location using public transport?
Q6	Are you willing to use at least 95% of all engineered wood products to meet the stipulated formaldehyde units? Or, are there no new engineered wood products used in the building?
Q7	Are you willing to source 95% of building steel from a responsible steel maker?
Q8	Are you willing to source 95% of timber used in the building with a forest certification scheme?
Q9	Are you willing to avoid using polyvinyl chloride (PVC) (at least 90%, by cost) and have an environmental product declaration (EPD)?
Q10	Are you willing to use an on-site renewable energy generation source?
Q11	Were there any critically endangered, or vulnerable species, or ecological communities present on the site at the time of purchase?
Q12	Was the ecological value improved by the project?
Q13	Is the site classified as a site of "High National Importance?"
Q14	Is the site a previously developed land (at least 75%)?
Q15	Is the land previously contaminated and will the site be remediated with a best practice remediation strategy?
Q16	Are you willing to reduce the use of Portland Cement content in all the concrete used in the building by replacing it with supplementary cementitious material (SCM)?
Q17	Are you willing to use 95% of all internally applied paints, adhesives, sealants and carpets to meet the stipulated "total volatile organic compound (VOC) levels, or not to use paints, adhesives, sealants and carpets?"
Q18	Does the building has a clear line of sight to a high quality internal or external view?
Q19	Is the building located conveniently vis-à-vis amenities?
Q20	What is the type of ventilation system used in the building?

identifies the constraint regarding the accessibility of the building site. This has a direct impact on credits "T17.B1 Access by Public Transport." If the location does not provide access by public transport, this credit should be excluded from the selection. Similarly, questions 11, 12, 13, 14, and 15, focus on the ecological value of the land. Question 11 seeks information to satisfy "L23.0 Endangered, Threatened or Vulnerable Species" credit, and question 12 focuses on "L23.1 Ecological Value" credit (refer Table 6.1). The aim of these credits is to reward projects that improve the ecological value of the site. Similarly, questions 13 and 14 seek information on whether the land was previously developed or not, and whether the land was of national importance or not. These two questions provide necessary information to credit "L24.1 Reuse of Land" and "L24.0 Conditional Requirement." Question 15 requires information on any contamination of the land and its remediation. This question directly relates to "L24.2 Contamination and Hazardous Material" credit. All these requirements depend on the plot of land and its location which is already been selected by the user. Therefore, the model initially obtained information on the constraints the user has regarding these credits. As an example, if the selected plot of land is located in a secluded area without access to public transport, it is impossible to achieve the "T17.B1 Access by Public Transport" credits. In such circumstances, the model identifies this user constraint and eliminates this credit instantly. Similarly, all these questions obtain information on the user constraints that occurs leading to elimination of credits.

Question 18 of the questionnaire focuses on the clear view of sight that the building has. This information directly links with "IEQ12.2 Views" credit. Further, question 19 provides information on the amenities close to the building. This question is related to "T17B.5 Walkable Neighborhood."

There are many questions related to material usage of the building. According to Green Star Design and As-Built version 1.1, there are certain materials that cannot be used in the building, certain material should be sourced from specific sources and the usage of certain materials needs to be reduced by using supplementary material. However, the users of this model, specifically the builders, might have specified procurement routes for construction materials. Therefore, the builders might not be in a position to easily procure from a given source, or sometimes, certain use of special material might require special skilled people. To cater to this requirement, this model asks a series of questions regarding construction material.

Questions 7 and 8 require information on the suppliers of steel and timber. The main reason for this is that Green Star Design and As-Built version 1.1 require the steel to be sourced from a responsible steel maker, and timber to be certified by product certification authorities. These questions are directly linked with credits "Mat20.1 Structural and Reinforcing Steel" and "Mat20.2 Timber Products." Similarly, questions 6 and 17 focuses on the specifications of certain material. Question 6 focuses on the use of engineered wood products and level of formaldehyde content in them. Further, the questions discuss the paints, adhesives, sealants and carpets and the levels of VOCs. Both these credits directly focus on "IEQ13.1 Paints, Adhesives, Sealants and Carpets" and "IEQ13.2 Engineered Wood products" respectively.

The usage of PVC in the building is examined through question 9. Most of the buildings use PVC products for pipes, cables and so on. Green Star Design and As-Built version 1.1, either require avoiding the use of PVC altogether, or having an EPD while using PVC. This information has a direct impact on the credit "Mat20.3 Permanent Formwork, Pipes, Flooring, Blinds and Cables." Similarly, question 16 focuses on the usage of SCM to reduce the use of cement in concrete. This question is supported by credit "Mat19B1.1 Portland Cement Reduction."

The willingness to use an on-site energy generation source is examined under question 10. The main consideration here for on-site energy generation is PV panels, as illustrated in Section 5.6. By using on-site energy generation, the total peak electricity demand can be reduced. Therefore, this question directly relates with "E16A Prescriptive pathway: On-site energy generation" credit.

Fig. 6.1 reports the input screen of the life-cycle cost model for green commercial buildings. The input screen includes all the questions as user inputs.

6.3.3 Exclusions, dependencies and selections based on user inputs

According to Section 6.3, there are two types of user inputs. There is information provided by the user, and the constraints. Based on the information provided by the user, the model makes certain selections. When the user provides the CBD where the building is located, the model selects the relevant cost data for the respective CBD. Afterwards, once the user provides the discount rate, the model re-calculates the necessary life-cycle costs. Based on the desired certification, the model defines the boundaries. Finally, if the building is located in a regional area, the user can specify the regional area, whereas the model re-adjusts the life-cycle cost values to account for regional changes.

For user constraints, the model runs an elimination process. As an example, if the user states that the land is not located in an area with access to public transport, the model eliminates the related credit, and in this instance, the credit is "T17.B1 Access by Public Transport." For this elimination process, the model uses the following rule given in Fig. 6.2.

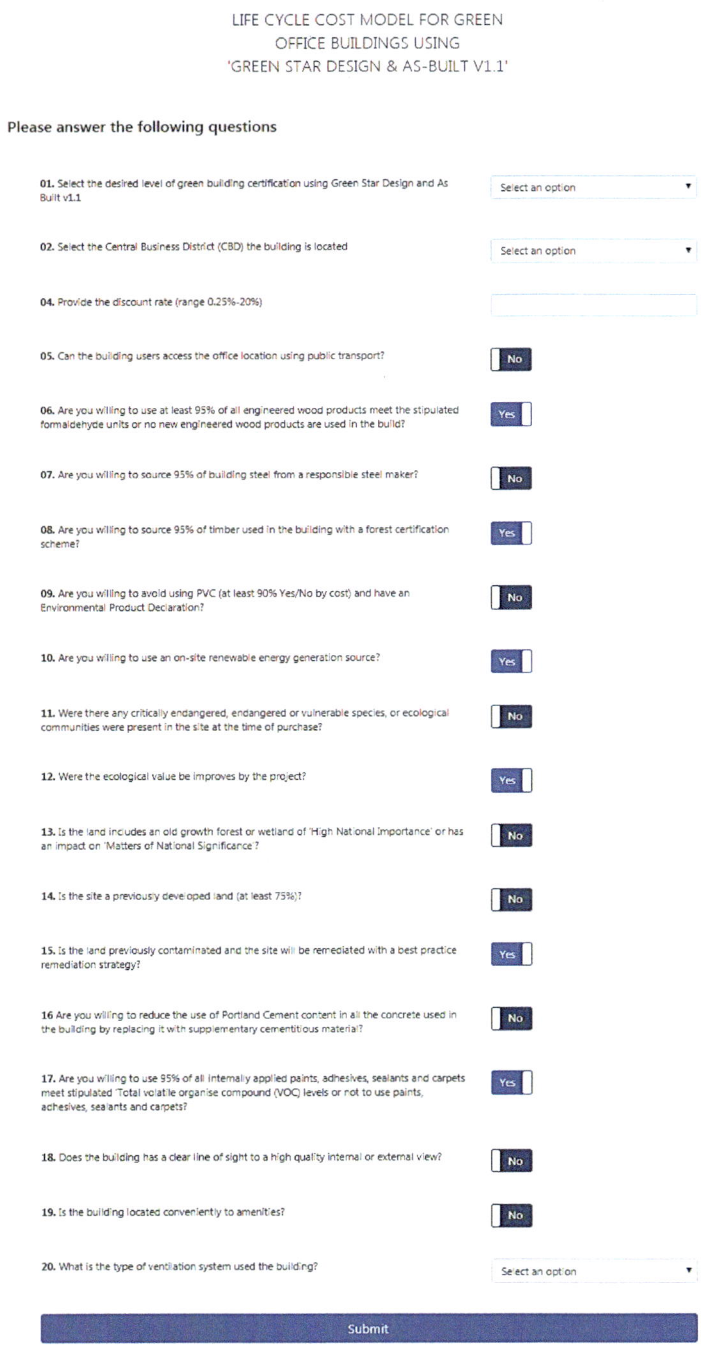

FIG. 6.1 User input screen of life-cycle cost model.

```
{
    "parameter": "q5-access-by-public-transport",
    "type": "direct",
    "rules": [
        {
            "exclusions": [
                "17b-1-access-by-public-transport"
            ],
            "value": "false"
        }
    ]
}
```


FIG. 6.2 Exclusion rule.

In the exclusion rule, the "parameter" refers to the question and the "value" refers to the answer from the user. If the user says "no" as the answer, which is given as "false" in the rule, the credit "T17.B1 Access by Public Transport" is eliminated for the selection rounds by the model (refer Fig. 6.2). Similarly, if the "value" is given as "true," then the particular credit is not eliminated for the selection rounds (Table 6.2).

The last question of the questionnaire is on the type of ventilation system of the building. There are three options available for this question, namely, naturally ventilated, mechanically ventilated with water cooled

TABLE 6.2 Credits eliminated from the exclusion rule.

User question number	Eliminated credit if the user input is "No"
Q5	T17.B1 Access by Public Transport
Q6	IEQ13.2 Engineered Wood products
Q7	Mat20.1 Structural and Reinforcing Steel
Q8	Mat20.2 Timber Products
Q9	Mat20.3 Permanent Formwork, Pipes, Flooring, Blinds and Cables
Q10	E16A Prescriptive Pathway: On-site Energy Generation
Q12	L23.1 Ecological Value
Q14	L24.1 Reuse of Land
Q15	L24.2 Contamination and Hazardous Material
Q16	Mat19B1.1 Portland Cement reduction
Q17	Mat20.3 Permanent Formwork, Pipes, Flooring, Blinds and Cables
Q18	IEQ12.2 Views
Q19	T17B.5 Walkable Neighborhood

```json
{
  "parameter": "q20-ventilation",
  "type": "direct",
  "rules": [
    {
      "exclusions": [
        "9-2-provision-of-outdoor-air-naturaly-ventilated",
        "9-2-provision-of-outdoor-air-mechanically-ventilated-air-cooled-system",
        "14-1-1-thermal-comfort-naturally-ventilated",
        "14-1-2-thermal-comfort-mechanically-ventilated-air-cooled-system",
        "14-2-advanced-thermal-comfort-naturally-ventilated",
        "14-2-advanced-thermal-comfort-mechnically-ventilated-air-cooled-system",
        "15a-ghg-emissions-reduction-naturaly-ventilated",
        "15a-ghg-emissions-reduction-mechanically-ventilated-air-cooled-system",
        "28-legionella-impacts-from-cooling-systems-naturally-ventilated",
        "28-legionella-impacts-from-cooling-systems-mechanically-ventilated-air-cooled-system",
        "29-refrigerant-impact-naturally-ventilated",
        "29-refrigerant-impact-mechanically-ventilated-air-cooled-system",
        "18b-3-heat-rejection-mechanically ventilated-air-cooled-system",
        "18b-3-heat-rejection-naturally-ventilated"
      ],
      "value": "mechanically-ventilated-water-cooled-sysstem"
    }
  ]
}
```

FIG. 6.3 Exclusion rule for mechanically ventilated system.

system and mechanically ventilated with air cooled system. According to Section 4.4, there are various inter-dependencies based on the types of ventilation. There are several credits that can be combined together to achieve desired total credit points. Further, based on the type of ventilation in the building, there are significant changes in the life-cycle cost to various credits (refer Section 5.6). Therefore, when the user specifies the type of ventilation in the building, the model eliminates various costs related to different credits calculated for other ventilation types. However, the rule changes slightly for naturally ventilated buildings and mechanically ventilated buildings. Fig. 6.3 illustrates the exclusion rule for mechanically ventilated buildings.

According to Fig. 6.3, the "parameter" refers to question 20, and if the "value" given, or in other words, the answer is mechanically ventilated water cooled system, then all the credits, together with the life-cycle cost relating to naturally ventilated buildings and mechanically ventilated with air cooled system, are eliminated by the model for the selection rounds. However, for naturally ventilated buildings, there are certain credits that can be directly attributed (refer Section 4.4). Therefore, the model directly includes these credits, while eliminating the credits related to mechanically ventilated buildings. Fig. 6.4 illustrates the exclusion rule for naturally ventilated buildings. According to Fig. 6.4, the model includes "Em28 Legionella impacts from cooling systems" and "Em29 Refrigerant Impact credit" directly included for selections.

Apart from the exclusion rules, there are another set of rules focusing on the inter-dependencies among the credits. According to Section 4.4 and

```
{
  "parameter": "q9-ventilation",
  "type": "direct",
  "rules": [
    {
      "inclusions": [
        "28-legionella-impacts-from-cooling-systems-naturally-ventilated",
        "29-refrigerant-impact-naturally-ventilated"
      ],
      "exclusions": [
        "9-2-provision-of-outdoor-air-mechanically-ventilated-water-cooled-system",
        "9-2-provision-of-outdoor-air-mechanically-ventilated-air-cooled-system",
        "14-1-2-thermal-comfort-mechanically-ventilated-water-cooled-system",
        "14-1-2-thermal-comfort-mechanically-ventilated-air-cooled-system",
        "14-2-advanced-thermal-comfort-mechnically-ventilated-water-cooled-system",
        "14-2-advanced-thermal-comfort-mechnically-ventilated-air-cooled-system",
        "15a-ghg-emissions-reduction-mechanically-ventilated-water-cooled-system",
        "15a-ghg-emissions-reduction-mechanically-ventilated-air-cooled-system",
        "28-legionella-impacts-from-cooling-systems-mechanically-ventilated-water-cooled-system",
        "28-legionella-impacts-from-cooling-systems-mechanically-ventilated-air-cooled-system",
        "29-refrigerant-impact-mechanically-ventilated-water-cooled-system",
        "29-refrigerant-impact-mechanically-ventilated-air-cooled-system",
        "18b-3-heat-rejection-mechanically-ventilated-air cooled system"
      ],
      "value": "naturally-ventilated"
    }
  ]
}
```

FIG. 6.4 Exclusion rule for naturally ventilated system.

Fig. 4.3 there are various dependencies among the credits. This model integrates all of these dependencies to arrive at an optimal solution.

Conditional credits are of great importance. While selecting the optimum credits, the model includes all the related conditional credits, as well. Conditional credits do not have any credit points. Therefore, in the selection rounds if the model selects any credit which falls into a sub criteria with a conditional credit, there is a rule developed to add the relevant conditional credit, as well. Fig. 6.5 illustrates the rule for conditional credits.

According to Fig. 6.5, if the model selects credit points from "M2.1 Services and Maintainability Review" credit, this rule identifies a dependency for the related conditional credit "M2.0 Environmental Performance Targets." Therefore, when the model selects "M2.1 Services and Maintainability Review" credit point in the selection rounds, it also

```
{
    "credit-point": "2-1-sevices-and-mainatainability-review",
    "dependencies": [
      "2-0-environmental-performance-targets"
    ]
  }
```

FIG. 6.5 Rule for conditional credits.

```
{
    "credit-point": "2-4-independent-commissioning-agent",
    "direct-dependencies": [
      "2-0-environmental-performance-targets"
    ],
    "conditional-dependiencies": [
      "2-1-sevices-and-mainatainability-review",
      "2-2-building-commissioning",
      "2-3-building-systems-tuning"
    ],
    "minimum-conditions" : "1"
  }
```

FIG. 6.6 Direct dependency rule.

adds the conditional credit as well. This rule is applicable to all the credits in sub criteria given in Table 4.1 in Section 4.3.1.

In Green Star Design and As-Built version 1.1, there are additional points available (refer Section 4.3.2). The model can only select an additional point after fulfilling the general credit. In such instances, a rule similar to the "rule for conditional credits" is used. This rule is applied for all the additional credits reported in Table 4.1.

There are certain direct dependencies given in the Green Star rating tool. As an example, the credit on "M2.4 Independent Commissioning Agent" can only be used if the any of the credits in that particular sub criteria is selected. Further, that subcriteria also has one conditional credit, as well. Therefore, in such instances, the direct dependency rule is used as reported in Fig. 6.6.

There are a lot of inter-dependencies among the credits as given in Section 4.4 and Fig. 4.3. A separate rule is applied for these inter-dependencies. Fig. 6.7 reports the rule for inter-dependencies.

According to Fig. 6.7, there is an interdependency between "L25 Heat Island Effect Reduction" and "E16A Prescriptive Pathway: On Site Energy Generation" (refer Section 4.4). The model collectively considers these credits with inter-dependencies. Therefore, each inter-dependent credit is included in this rule. Once the model consider all these rules, it starts the selection process.

```
{
    "credit-point": "25-1-heat-island-effect",
    "inter-dependencies": [
      "16a-prescriptive-pathway-on-site-energy-generation"
    ]
  },
```

FIG. 6.7 Rule for inter-dependencies.

6.4 Life-cycle cost model development

The initial model involves identifying the user constraints and inputs. Based on the inputs, the life-cycle costs change, and finally, the model selects the optimized solution with the lowest life-cycle cost, and the maximum credit points for the selected certification level.

The main user information included in the model is the discount rate, project location, and certification level required. Based on the certification level, the model sets lower and upper boundaries for the calculation of total credit points. The model receives user constraints in the form of responses to a questionnaire (refer Section 6.3).

According to Section 2.4 in literature, green buildings should satisfy the environmental, social and economic sustainability norms, which are represented by key criteria and credits of green building rating tools. However, a building may be certified as a green building even if one of the parameters, such as "water," is completely ignored. As an example, in Green Star Design and As-Built version 1.1, the total energy credits amounts to 22 credits points (if any of the credits are not eliminated), IEQ amounts to 17 credit points and land use and ecology credits amount to a total of six credit points. Therefore, a building can achieve all these credits focusing only on the three key criteria and achieve a total of 45 credit points in total, fulfilling the four-star Green Star certification. However, in this scenario, other key criteria such as management, transport, water, material and emissions are completely ignored. Therefore, such a scenario is not encouraged by this model; thus, it implements a rule to include at least one credit from each key criterion.

Fig. 6.8 below shows the life-cycle cost model algorithm.

Initially, the model eliminates the credits with constraints (refer Section 6.3). This model has two main rounds of selections (refer Fig. 6.8). The first round selects the optimum credits from each key criterion. Once this selection is done, the model checks whether there are any conditional credits need to be added. If so, the model adds the conditional credits. Afterwards, the model runs for the second round selections. In this selection, the model identifies the lowest life-cycle credits with the highest credit points and adds them to the selection list. With each selection, the model, once again, checks whether any conditional credits need to be added or not. All the selections are based on the rules reported in Section 6.3.3. The model initially fulfills the lower boundary of credit points. Once it is fulfilled, then, it iteratively selects the credits for the middle and the upper boundary. However, if there are many user constraints, then the model might not be able to reach the boundaries stipulated by the user. If the score is greater than the minimum, the model presents all possible options. However, if the score is less than the required minimum, the model identifies the lowest life-cycle option eliminated owing to the

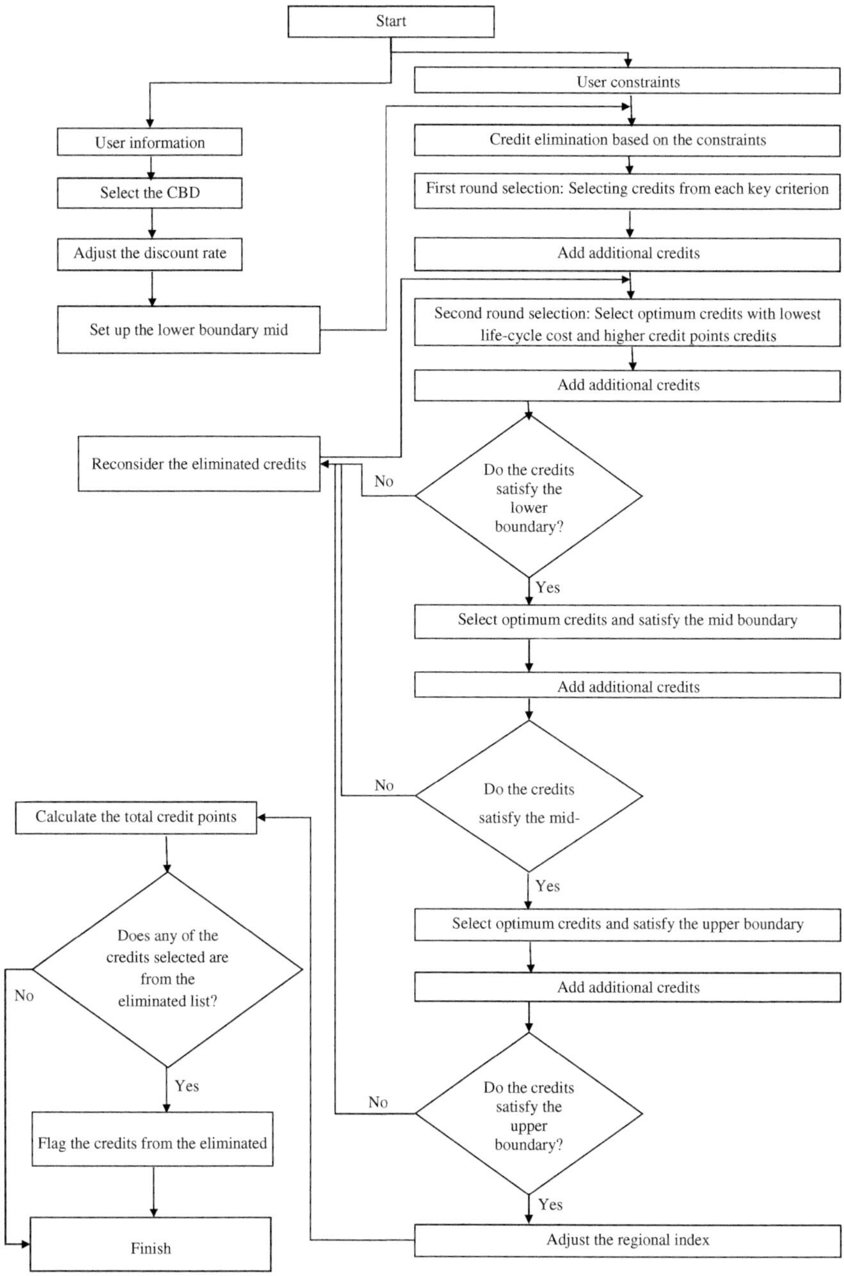

FIG. 6.8 Algorithm for the life-cycle cost model.

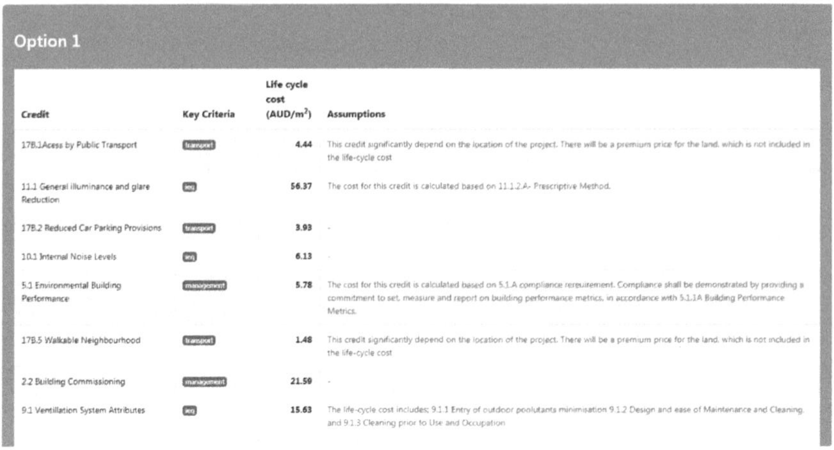

Option 1

Credit	Key Criteria	Life cycle cost (AUD/m²)	Assumptions
17B.1Acess by Public Transport	transport	4.44	This credit significantly depend on the location of the project. There will be a premium price for the land, which is not included in the life-cycle cost
11.1 General illuminance and glare Reduction	ieq	56.37	The cost for this credit is calculated based on 11.1.2.A- Prescriptive Method.
17B.2 Reduced Car Parking Provisions	transport	3.93	-
10.1 Internal Noise Levels	ieq	6.13	-
5.1 Environmental Building Performance	management	5.78	The cost for this credit is calculated based on 5.1.A compliance rereuirement. Compliance shall be demonstrated by providing a commitment to set, measure and report on building performance metrics, in accordance with 5.1.1A Building Performance Metrics.
17B.5 Walkable Neighbourhood	transport	1.48	This credit significantly depend on the location of the project. There will be a premium price for the land, which is not included in the life-cycle cost
2.2 Building Commissioning	management	21.59	-
9.1 Ventillation System Attributes	ieq	15.63	The life-cycle cost includes; 9.1.1 Entry of outdoor poolutants minimisation 9.1.2 Design and ease of Maintenance and Cleaning and 9.1.3 Cleaning prior to Use and Occupation

FIG. 6.9 Final output of life-cycle cost model.

constraints and provides these options by presenting the next best ones (refer Fig. 6.8). These credit points with constraints are presented to the user as red flagged credits, so that the user can consider eliminating the constraints if the project needs higher levels of certification.

The user interface of the model provides three options as solutions. Fig. 6.9 illustrates the final output of the model. In the output screen, the user can visualize the three options and when the user clicks on each option, it provides further details on each option including the credit points, key criteria, life-cycle cost in AUD/m^2 and the assumptions made for calculation (refer Fig. 6.9).

The life-cycle cost model operates in an external server, and therefore, it can be accessed through a web link. The link to access the life-cycle cost model is http://www.outreech.net/c/. Apart from that, Section 6.4.1 provides a user guide to help understand the proposed model. All the life-cycle cost data is uploaded onto Google Documents. Therefore, the life-cycle cost data and regional indices can be updated regularly.

6.4.1 User guide to the proposed life-cycle cost model

As illustrated in Section 6.3, the user has to input the user information and user constraints to the model in the input screen. Once the user submits the form, the solutions appear in the output screen. If the user fails to provide any user information in the input screen, an error message pops-up indicating the missed information. Further, when the model selects an eliminated credits due to user constraints, (refer Section 6.4), a notifications pops-up. Following steps illustrates the steps in using the proposed life-cycle cost model.

Step 1: The user should click on the following link or copy and paste in the browser to open the life-cycle cost model:
http://www.outreech.net/c/

Once the user follow the link the input screen will appear. It includes a series of questions to obtain user inputs and constraints (refer Section 6.4). The user has to answer the series of questions to provide information.

Step 2: The user should provide the user inputs selecting from the dropdown lists. As illustrated in Section 6.4, the first four questions of the input screen obtain the user inputs. Initially, the user should specify the level of certification required for the green building. Fig. 6.10, illustrates the user input screen for certification level.

Once the level of certification is identified, the user should specify the closest CBD. This proposed life-cycle cost model obtained cost data from six CBDs (refer Chapter 5), and therefore, this model can tailor made the solutions to suit a specific CBD. Fig. 6.11, illustrates the input screen for selecting the CBD.

Based on the selected CBD, the user can then select the relevant regional area. A dropdown list appears as given in Fig. 6.12 to select the relevant regional area. If the green building is located in the CBD, the user can select the name of the CBD in the list.

Step 3: User should type the discount rate in the given space.

As explained in Section 5.2, users have their own discount rates. Therefore, this proposed life-cycle cost model allows the users to input the discount rate and calculates the life-cycle cost based on the discount rate given by the user. Fig. 6.13, illustrates the user input on discount rate.

If the user fails to provide any of the user inputs and submit the form, an error message pops-up as illustrated in Fig. 6.14.

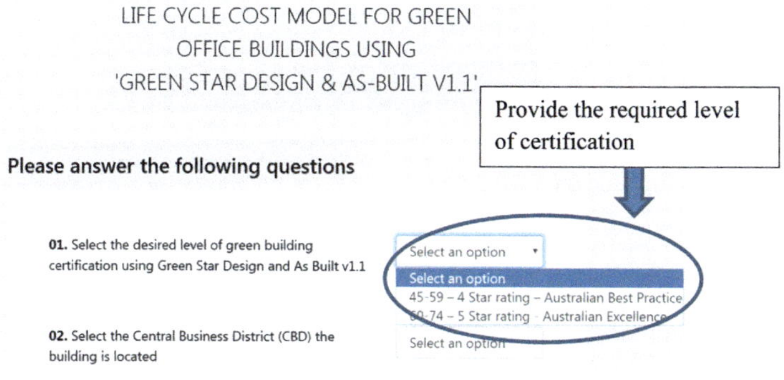

FIG. 6.10 Information on level of certification required.

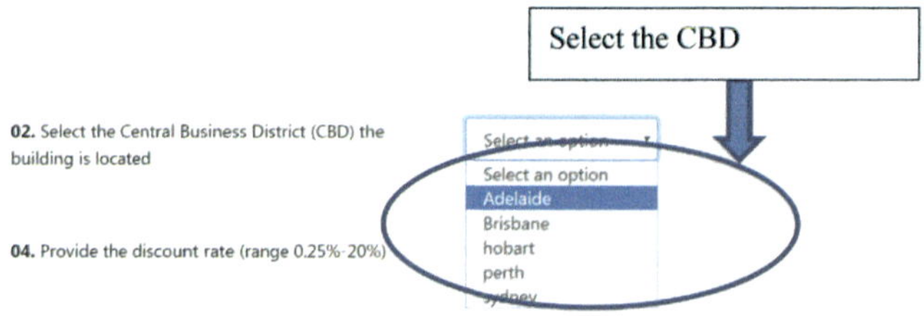

FIG. 6.11 Information on the CBD.

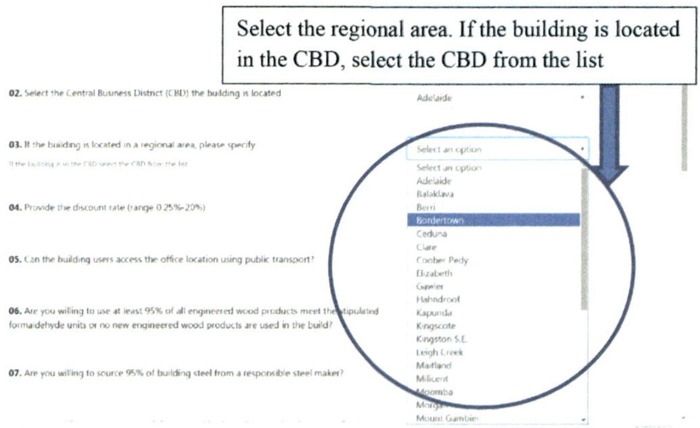

FIG. 6.12 Information on regional areas.

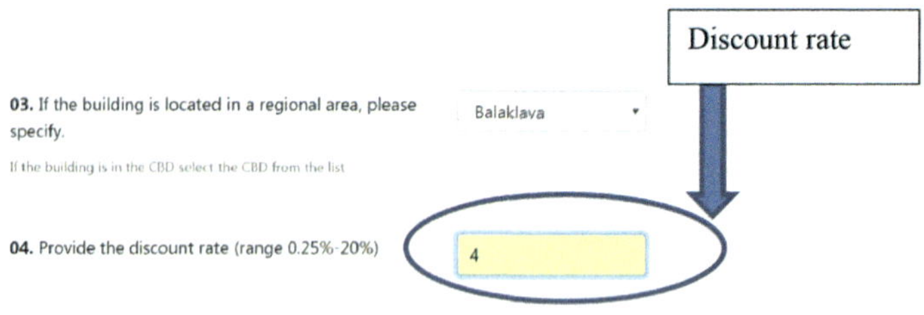

FIG. 6.13 Information on discount rate.

FIG. 6.14 Error message to notify missing information.

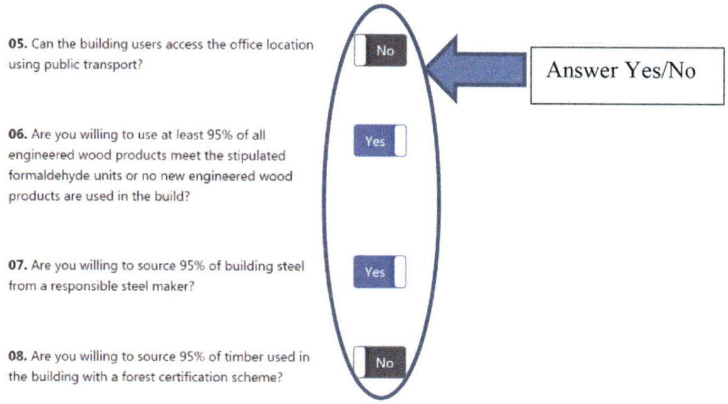

FIG. 6.15 Information on discount rate.

Step 4: The user should answer the question Yes/No from question 5 to 19. As illustrated in Section 6.4, these questions represent the user constraints. Fig.6.15, illustrates the input screen for user constraints.
Step 5: The user should select the type of ventilation from the dropdown list. The type of ventilation system installed in the building has many interdependencies (refer Section 6.3.3). Therefore, this model gets information on the proposed ventilation system for the building. Fig. 6.16, illustrates the input screen for type of ventilation system in the proposed life-cycle cost model.
Step 6: Once all the information is given, the user can submit the form. Once the user click "Submit," all the solutions will appear on the screen.
If any of the options includes credits which are eliminated due to user constraints (refer Section 6.4), the proposed model notify that by giving a pop-up as illustrated in Fig. 6.17. The user can identify these constraints and eliminate these to achieve higher green building ratings.

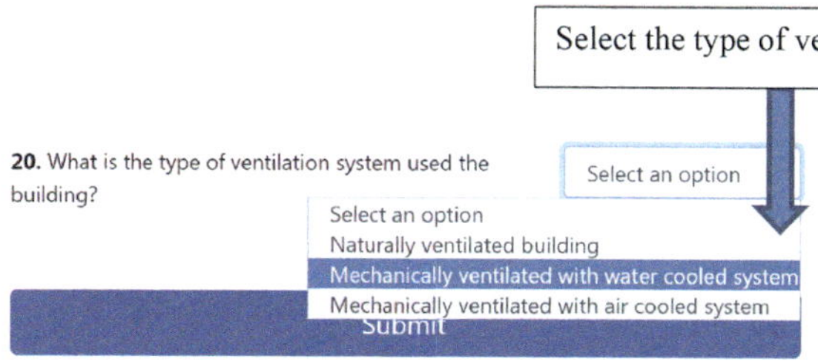

FIG. 6.16 Information on the type of ventilation system.

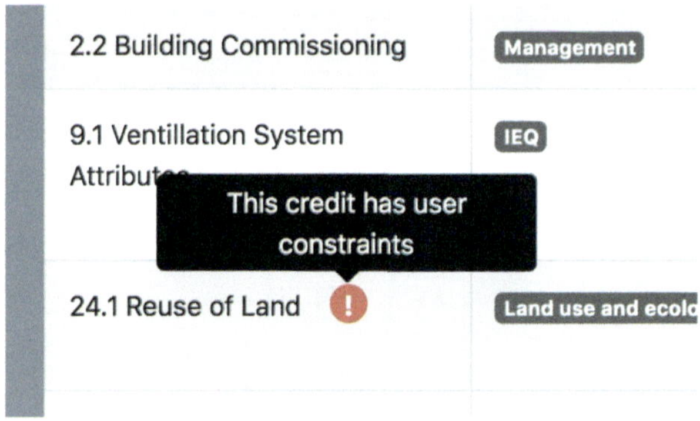

FIG. 6.17 Message on user constraints.

6.5 Changes to the cost model considering widely used international green building rating tools

There are many green building rating tools used worldwide. In Section 2.4.1 of this book, authors identified eight green building rating tools (including Green Star) widely used. Most of these green building rating tools follow a similar credit point system to evaluate green buildings and as users can develop a similar life-cycle cost model. Green buildings rating tools such as LEED, BREEAM, GBI, Green Mark, IGBC and BEAM Plus follow a similar scoring system although the number of points given for credits and credit classification are different. Therefore, similar life-cycle cost model can be adopted easily using the same methods. CASBEE gas a different scoring model for green buildings. CASBEE calculates "Built environment quality" opposed to "Built environment load" to evaluate the green building performance. Therefore, this life-cycle cost model cannot be directly adopted to CASBEE rating tool. Herein after this section refer to the above given green building rating tools except for CASBEE.

According to Table 2.1 in Chapter 2, all the green building rating tools have similar key criteria except for CASBEE. LEED has eight key criteria. Similar to Green Star, each of these key criteria are divided into credits and each credit has credit points. LEED also has pre-requisites. Those coincide with the "conditional credits" in Green Star. LEED also specifies various pathways to achieve certain credits and therefore, users must identify the pathways prior to developing the life-cycle cost model (refer Section 4.3.3). "Credits based on types of buildings," "Credits linked to another assessment too" and "Credits with choices" are also available in LEED. Therefore, prior to developing the life-cycle cost model, users must classify those credits as given in Section 4.3 of Chapter 4 of this book. The same process can be adopted for the other tools such as BREEAM, GBI, Green Mark, IGBC, and BEAM Plus. However, for each of these rating tools, users must initially analyze the credits and classify those prior to life-cycle cost calculation.

Upon completing the credits classification, calculation of the life-cycle cost is the next step in the process. Users can use the NPV technique to calculate the life-cycle cost model, yet certain formula should be updated to suite the geographic conditions. The user inputs are different for each green building rating tool. For example, in the user inputs, this model asks whether the users are willing to use at least 95% of all engineered wood (Q6 in Table 6.1). This question is based on the Green Star requirements. Although the key criteria of most of the green building rating tools are similar, the requirements given in the credits are different to suit the geographical conditions. Therefore, when developing a life-cycle cost model user must analyze the credits and develop the "user inputs" for each green building rating tool separately.

Users can use the algorithm used in Fig. 6.8 as a guide for developing the life-cycle cost model. In this life-cycle cost model authors used atleast one credit from each criterion. Green building rating tools such as BREEAM, defines minimum BREEAM standard. "Minimum BREEAM standards by rating level" refers to a set of credits identify as minimum requirements to achieve a given rating level (Building Research Establishment Environment Assessment Method, 2014). In such instances users can further develop the algorithm to cater these minimum requirements.

6.6 Summary

This chapter discussed the life-cycle cost model development. This chapter started by providing a brief idea on the life-cycle cost model. Then, it illustrated user inputs. There are two types of user inputs, namely, user information and constraints. User information provides data to calculate the specific life-cycle costs, and also provides data to set up the boundaries for certification. User constrains are critical in the optimum selection of credit points, because, based on the constraints, the model eliminates certain credits. Afterwards, the model selects the credits to satisfy the boundaries, considering the different selection rules, which have been illustrated in detail through this chapter. Finally, the model presents the final output with three options. Each option is illustrated in detail. If the user cannot achieve the desired green certification with the given certification parameters, this model also provides alternative options, so that the user can eliminate the optimum constraints in achieving the desired certification.

References

Building Research Establishment Environment Assessment Method. (2014). *BREEAM international new construction technical manual (SD 5075 1.0:2013)*. Retrieved from https://www.breeam.com/.

Currie, E. (2006). *Fundamentals of programming using Java*. Cengage Learning EMEA.
Naressi, A., Couturier, C., Devos, J., Janssen, M., Mangeat, C., De Beer, R., & Graveron-Demilly, D. (2001). Java-based graphical user interface for the MRUI quantitation package. *Magnetic Resonance Materials in Physics, Biology and Medicine, 12*(2–3), 141.
Rawlinsons. (2016). *Construction cost guide for housing, small construction and industrial buildings. Vol. 34*. Perth, Western Australia: Rawlinson Publishing.

CHAPTER

7

Life-cycle cost model validation

7.1 Introduction

This chapter discusses the validation of the life-cycle cost model. However, this validation is twofold. It includes validations of the cost data and validation of the credits selection of the life-cycle cost model using four case studies. The cost data are validated using the cost database composite rates and discussions with estimators. This chapter extensively illustrates each of these validation methods used.

7.2 Validation of costs

This life-cycle cost model involves a massive set of cost calculations. The accuracy of the selections depends on the accuracy of the cost estimates. However, the costs change overtime and therefore, these costs need to be updated regularly. The cost data for this model are uploaded as an Excel sheet in Google Drive. Therefore, the cost data can be directly updated in regular intervals so that the life-cycle cost model uses the most recent cost data to select the optimum solutions. By updating the cost data regularly, this life-cycle cost model tends to provide accurate cost data to the users.

Further, as illustrated in Chapter 5, the life-cycle cost data are developed based on the first principles. This includes using various sensible assumptions. These assumptions are clearly stated in the output of the life-cycle cost model. Therefore, the user can make better decisions considering the assumptions. This helps to enhance the accuracy of the model as well. Apart from that, there are mainly two methods used to validate the accuracy of cost data. These methods are verification by using cost databases and cross checking with industry experts.

The proposed model's accuracy was verified in terms of life-cycle costs and credit selection. Initially, all life-cycle cost data were verified by three professional estimators. This study always focused on obtaining industry inputs

https://doi.org/10.1016/B978-0-12-820062-9.00007-4

TABLE 7.1 Details of estimators.

Estimator	Designation	Years of experience (years)
Estimator 1	Senior Quantity Surveyor	12
Estimator 2	Quantity Surveyor (Mechanical, Electrical, and Plumbing)	8
Estimator 3	Quantity surveyor	5

from industry professionals regarding the cost calculations. The cost calculations were constantly reviewed and discussed among estimators working in the Australian construction industry. There were three estimators reviewing the cost data throughout the study. Details of the professionals are reported below in Table 7.1. All the estimators have experience in working in green building projects. Estimator 2, (refer Table 7.1) is currently working in a green buildings project in Australia. Estimator 1 has more than 3 years of experience in working in green building projects especially in green office buildings.

All the estimators had years of experience and the cost calculations were discussed regularly with the team of estimators. If any cost was unrealistic or inconsistent, the costs were reviewed and adjusted to ensure the accuracy of the costs.

Apart from the verification from industry experts, the cost calculations were cross-checked against databases, reports, and an actual Green Star certified building. The Green Building Council of Australia (2016) showed that the cost of a four- or five-star-rated building ranges from 3020 to 3536 AUD/m^2. The sum of the initial cost per square meter of a building in Sydney with air-conditioning and a five-star rating is approximately 3250 AUD/m^2; therefore, it is within the appropriate range. Furthermore, this initial cost included the costs of obtaining the green credits; therefore, certain incidental items are omitted from the cost. There are many options available for obtaining a green star rating. However, to arrive at this figure, a total of 63 credits were selected from the model.

Similarly, the Cordell (2016) database for commercial buildings in New South Wales (NSW), stated that the unit rate for an average concrete-framed office building including air-conditioning is approximately 2418 AUD/m^2 in NSW. However, according to the Green Building Council of Australia (2016), on average the developers achieve a Green Star rating with a 3% increment. Therefore, if the initial cost for a conventional building is 2418 AUD/m^2, the cost of a similar Green Star rated building would be approximately 3150 AUD/m^2. This approximate cost figure is very similar to the figure obtained from the cost calculation. According to Rawlinsons (2016), a fully serviced office building in the Sydney area ranges from 2285 to 2465 AUD/m^2. Once again, considering the green cost premium, the green building cost for an office building varies from 2970 to 3204 AUD/m^2. In this instance also, the cost per square meter of a green office building area is closely similar to the calculated figure in the cost calculations of this study. These cost figures are computed based on various sensible assumptions and factors. Therefore, it is highly unlikely that that two total cost estimates for the same building will have the same values. However, the cost figures should be somewhat similar, which is satisfied in all these cases.

The Australian Institute of Quantity Surveyors (2017) published that an office building in Sydney CBD with air-conditioning and standard finish would cost approximately 2790 AUD/m^2. If the 3% premium is added, the cost would be 3627 AUD/m^2. However, the costs given by the Australian Institute of Quantity Surveyors (2017) included the profits. Therefore, this inclusion explains the increase in the per square meter cost compared to the costs given in other databases. The green office building developed by the cost data published by the Australian Institute of Quantity Surveyors (2017) is slightly higher than the calculated cost figure. The main reason is the profit adjustment, where as in the cost calculations for life-cycle cost model excluded the profit adjustments.

The building cost for the Council House 2 (CH2) office building in Melbourne is AUD 51.01 million (Green Building Council of Australia, 2008). This building was awarded a six-star Green Star rating. The total cost included AUD 2.8 million on education and demonstration and AUD 7.1 million for specific Council requirements. Therefore, the actual building costs amount to AUD 32.2 million. The gross floor area of the building is 12,536 m^2. The cost per square meter for the CH2 building is AUD 3287 AUD/m^2. The cost for a similar building in Melbourne CBD is approximately 3230 AUD/m^2 based on the cost calculation. Once again, the CH2 building cost is inclusive of profits. However, the cost calculated by the life-cycle cost model is exclusive of profit. This explains the main difference in the cost estimate.

All the cost figures on the cost of Green Star office buildings are consistent with the costs calculated for the study. There are comparisons between three Australian cost databases and the report published by the Green Building Council of Australia after reviewing the Green Star certified buildings within Australia. Further, the cost is compared against that of an actual Green Star certified office building as well. In all these cases, the costs calculated for the life-cycle cost model fall within the given cost ranges. Therefore, the cost estimates are consistent and accurate.

7.3 Validation of credit selection

According to Yin (1994, p. 14), a case study is an empirical inquiry that investigates a contemporary phenomenon within its real environment. In this study, real life Green Star certified green buildings need to be tested against the proposed model to evaluate the accuracy of the credit selections. There are many Green Star certified commercial office buildings in Australia. Therefore, prior to the selection, a clear method should be identified to act a basis for the case study selection.

There is a question as to how many case studies to select. However, according to Fellows (2015), this depends on the purpose of the study and the nature of the case studies. In this study, the purpose of using the case studies is to test the proposed life-cycle cost model. Further, the inputs of each case study differ, providing a novel scenario to the proposed model. Therefore, to test the accuracy of the proposed model, there should be various case studies presented, providing different input parameters. Considering the inputs, this study initially identified case studies from different CBDs to allow for changes in location. Afterwards, the case studies were selected based on the user inputs. Case studies are selected to reflect the user inputs and to present different scenarios of user inputs.

Due to the number of credits considered, the study only considered up to a five-star rating with total credit points ranges from 60 to 74. However, when selecting case studies, the authors selected case studies to satisfy a six-star rating with total credit points exceeding 75. The main reason for this selection is to identify the credits that are undertaken to reach total credits points beyond 75.

For the purpose of comparing results, for all the case studies, the discount rate is identified as 3.25% in the proposed model. Considering all the user inputs, authors selected four case studies as follows:

- Case study A—City central tower in South Australia
- Case study B—Council House 2 (CH2) building in Melbourne
- Case study C—Workplace 6 office building in Sydney
- Case study D—Bishop See South Tower in Perth.

When all the user inputs were included in the proposed model, it provided three options for the given certification level. From the three available options, the mid-range credit points are compared against the actual initiatives undertaken by each of these case study buildings to achieve the desired Green Star rating.

The model considered four case study office buildings that fulfilled all the constraints and compared the credits for certification attained by the building with the credits selected by the proposed life-cycle cost model. The following sections provide details of the case studies and details of the underlying constraints and user inputs. All the case study buildings are selected from the collection of case studies published by the Green Building Council of Australia (2008).

7.3.1 Case study A—City central tower of South Australia

The Case Study A building is the City Central Tower of South Australia, which is a new commercial office building with a five-star rating. It is one of the largest Green Star developments, with a net lettable area of over 30,000 m^2. Numerous initiatives are undertaken to obtain star ratings, and these initiatives cater to all the main key criteria of the building.

This building achieved over 90% of the management credits, including building commissioning and tuning, use of an Independent Commissioning Agent, metering, and provision of a building user guide. The contractor has an environmental management plan (EMP) and is accredited under the ISO 14001 Environmental Management System, which has ensured the environmental management of the construction site.

The City Central Tower provided exceptional IEQ standards, gaining a majority of IEQ credit points. The building's ventilation rate is a 100% improvement on the Australian standard, and the building used a 100% fresh air supply without any re-circulated component. Further, the building used electric lighting levels and high frequency ballasts that improve occupant comfort, including T5 fluorescent lighting with occupant control with dimmable ballasts. Apart from that, the building provided high thermal comfort, effective acoustic comfort, and low levels of indoor air pollutants by using low VOC carpets, adhesives, sealants, and composite wood products. There is a dedicated exhaust riser to remove indoor pollutants from printing and photocopy areas. Further, the building used a highly efficient spectrally selective façade glazing with external shading.

This building has a five-star Australian Building Greenhouse Rating (ABGR) certification, currently known as NABERS rating. Therefore, the building scored up to 16 credit points from this certification by using the "15D GHG Emissions Reduction-NABERS Energy Commitment Agreement." However, this credit is eliminated from this study (refer Section 4.3.8).

The City Central Tower was designed to reduce car parking allowance by 80% to achieve significant carbon dioxide reduction by removing approximately 770 cars from the road. Further, this building is located within close proximity to public transport and amenities. The land is re-used from an under-used city center site. The building also utilizes efficient sanitary fixtures, such as waterless urinals and taps with flow restrictors. There is a water-efficient irrigation system for the building.

During the construction phase of the building, 60% of construction waste was diverted from landfills. Further, the building provided recycling facilities for office waste within the operational phase of the building life-cycle. The timber used is sustainably sourced and 80% of the steel used in the building is of a 100% post-consumer content. The refrigerants and thermal insulation used are with a zero ozone depleting potential (ODP). There is storm water pollution management and treatment, and efficient water fittings reduce the flow to sewer. There are no external upward lights dispersed from the building to the external environment.

This building achieved credit points through "innovation." This key criterion is not considered in this study (refer Section 4.3), and it included credit points for the inclusion of direct tower cooling, enabling the building to be efficiently cooled directly by chilled beams circulating water through the cooling towers.

7.3.2 Case study B—CH2 building in Melbourne

CH2 is a 10-storey office building with a GFA of 12,536 m^2. It is one of the initial six-star certified commercial office buildings in Australia. The total cost of the building was AUD 51.045 million. The main focus of the building is to provide a healthy and productive workplace for its occupants, while reducing its impact on the environment through excellence in design and innovation. There are many green initiatives undertaken to achieve this aim of the building development.

This building provided a building user guide for future building occupants. Within the construction phase, there was independent building commissioning and tuning carried out. Further, the building utilized best practice environmental management and waste, including management systems implemented for the construction phase. There are recycling facilities installed in the building for office waste recycling within the operational phase of the building life-cycle.

The CH2 building focuses significantly on the IEQ key criterion. There are many novel and effective initiatives deployed to provide better IEQ to the occupants. The building has a displacement ventilation system for fresh air delivery, including a 100% fresh air supply with no recirculated air. Therefore, this increased fresh air supply quantities to three times the Australian standard. Further, it included occupant controlled air vents. The building provided better air quality for the occupants by accommodating low levels of indoor pollutants using carpets, adhesives, sealants, and composite wood products with low VOC levels. The CH2 building also provided a high thermal comfort performance. Glare control is available in the building via shading which moves and responds to the sun, and 80% of office occupants have access to outside views. The building used a T5 lighting system with small area zoning and daylight responsive light dimming.

There is an 87% reduction in GHG emissions compared to a reference building, and the building is also certified by ABGR with five-stars, which is similar to NABERS. Therefore, the CH2 building obtained 16 credit points from the "15D GHG Emissions Reduction-NABERS Energy Commitment Agreement." However, this credit is eliminated from this study (refer Section 4.3.8). The building utilized several on-site energy generation sources, such as solar photovoltaic cells for electricity generation and wind turbines integrated in the building. Hot water for the building is provided through solar panels. There is a phase change material (PCM) thermal storage in the building and the building has a low energy cooling system via chilled ceiling and shower towers for cooling. The building uses night time cooling via natural ventilation.

This building is located at a location with ample public transport facilities for occupants and provides access to amenities as well. Parking is provided for bicycles, and 25% of parking accommodates small cars. There are cyclist showers and changing facilities to encourage occupants to use bicycles for commuting. The land for this building was previously used as a car parking lot. Therefore, there is a change in the surface of the land to both horizontal and vertical gardens from an impermeable concreted surface, providing an improvement in ecological value.

Additionally, the CH2 building used efficient water fixtures. It also used reclaimed water for the sprinkler system and included facilities for rainwater collection. By incorporating these initiatives, the building experiences a 72% reduction in mains water consumption. Further, this building reduces sewer emissions by 80% through multi-water reuse (MWR) plant. It also incorporates storm water pollution management as well. All the timber used in the building is from sustainable sources. The use of PVC in the building is minimal. Further, all the refrigerants are with of a zero ODP and installed with refrigerant leak detection system.

This building achieved additional credit points through the "innovation" key criterion, which is not included in this study (refer Section 4.3). A couple of initiatives were considered for this key criterion, such as the use of chilled ceilings, a MWR sewer mining plant, sprinkler water reclaim, PCM thermal storage, a shower tower for cooling, and building integrated wind turbines.

7.3.3 Case study C—Workplace6 in Sydney

Workplace6 is a waterfront Sydney development comprising of 18,000 m^2 with 6 levels. It was one of the initial six-star certified office buildings in New South Wales. The main focus of the building design is a reduction in carbon emission and water consumption. According to the Green Building Council of Australia (2008), this building only produces carbon emissions equivalent to 138 cars on a road, compared to those of 356 cars in a conventional building, and reduces water consumption to one Olympic-sized swimming pool, whereas a non-green green building consumes water that fills 14 Olympic-sized swimming pools per year.

In the process of certifying for a Green Star, this building achieved the majority of the management credits. The initial stages of the building received the service of a Green Star accredited professional and also a commissioning agent was engaged at the early state of the project to ensure the required commissioning and tuning. The building is installed with metering and monitoring. The contractor who worked for this project construction had an environmental plan and was registered for ISO 14001 certification. The building also provided a user guide. During its construction period, the building diverted 80% of the waste from a landfill.

The Workplace6 building provided better IEQ to its occupants. It used paints, carpets, adhesives, and sealants with low VOC levels. Further, the building always received natural daylight and provided better external views. There is a good lighting design within the building, with high frequency ballast, and the building used chilled beams for thermal comfort. Further, the building has greater fresh air rates than a reference building.

This building achieved 16 points for the Green Star rating by obtaining a five-star NABERS energy rating. These credits are obtained by achieving the "15D GHG Emissions Reduction-NABERS Energy Commitment Agreement" pathway. However, this credit is eliminated from this study (refer Section 4.3.8). Energy is co-generated using renewable energy sources, and this initiative reduces the peak energy demand.

The Workplace6 building is located closer to public transport and to other amenities. There are limited car parking places in the building although it provides extensive cyclist facilities. The land used for the building is a previously contaminated land. Therefore, this project eradicated the contamination and used the land for the construction of the building.

There are water efficient sanitary fixtures installed in the building and four-star water fixtures and six-star urinals. The blackwater recycling system installed in the building produces 45,000 L of gray water per day. Further, this building's premises use a water-efficient irrigation system. During construction, the building used recycled content and also recycled steel. All the timber products were sourced from sustainable forests. Further, the building used the strategies necessary to minimize the use of PVC. It used various initiatives to reduce the flow of gray water to the sewer. Apart from that, there is minimum light pollution from the building and a refrigerant leak detection system in place.

When arriving at the six-star certification level, Workplace6 obtained several credits in "innovation" key criteria, which are not considered in this study. These credits included initiatives such as a blackwater recycling system and the use of SCM in post-tensioned concrete slabs. Further, this building performs over and above the set energy benchmarks achieving credit points through the "innovation" key criterion for exceeding energy benchmarks.

7.3.4 Case study D—The Bishops see south tower in Perth

The Bishop See South Tower is a nine-level commercial office tower with a five-star Green Star certification. There are exceptional features embedded into the building's façade, which greatly contributed to the achievement of the project's Green Star rating. Features of the façade include full-height floor-to-ceiling high performance double glazing and effective external shading which prevents excessive solar heat gains and glare whilst maintaining high internal daylighting levels.

From the initial stages of the building, the building obtained the professional service of a Green Star Accredited professional on the design team. Further, this building followed the necessary building commissioning and tuning procedures. There is an independent building commissioning agent appointed for the building project. The head contractor of the building construction is ISO 14001 accredited and followed an extensive environmental management plan. Within the construction process, 60% of the construction waste is re-used and recycled.

The air conditioning system installed in the building increases outside air rates and provides a better air change effectiveness through the use of high induction supply swirl diffusers. There is a carbon dioxide monitoring and controlling system installed in the system as well. The building also maintained the required level of daylight factor, used high frequency ballast to avoid low frequency flicker of light fittings, and always maintained proper lighting levels. This also involved using effective external glazing to reduce eyestrain and provided visual connection to the outdoor environment. There is excellent acoustic control within the building and indoor air quality was also considered by using paints, carpets, adhesives/sealants with low VOC content. Further, the building used low emission formaldehyde composite wood products and there is also a dedicated floor-by-floor horizontal discharge tenant exhaust system to minimize the indoor pollutants levels.

This is a 4.5 star-rated building using the ABGR assessment, currently known as NABERS. By using the ABGR certification, the Bishop See building achieved 16 credit points from the "15D GHG Emissions Reduction-NABERS Energy Commitment Agreement" pathway which is eliminated from this study. Further, the building used high efficiency water-cooled, oil-free, magnetic bearing chillers, coupled with oversized cooling towers and a dry-cooler tenant supplementary air conditioning system with a heat exchanger to base-build cooling towers. This initiative saved water equivalent to 65 domestic swimming pools per annum. High efficiency T5 lighting is available throughout the building.

The Bishop See building reduced the number of car bays by 25% from the total local planning allowance, and 26% of total car-parking spaces are provided for small cars. Cyclist facilities are extensively provided and so are excellent local public transport facilities. The land of the building was previously a built-up site. However, there is no change in the ecological value of the site.

The building has one of the initial commercial gray water systems in Perth, using waste water from showers and sinks to flush toilets. There are water efficient fixtures and an irrigations system included in the building. The test water for the fire system is re-used. There is a dedicated recycling waste storage area in the building. The building uses sustainable timber and 76% of structural steel has post-consumer recycled content.

There is a refrigerant leak detection and recovery system used in the building and 100% of the HVAC refrigerants have zero ODP. There is a reduction in the flow of waste water to sewer due to gray water system used. The building does not create light pollution from external lighting beyond the site boundary.

7.3.5 User inputs for each case study

All the above sections provided a brief description of each of the four case studies. Based on the information collected, all the user constraints and information are fed to the proposed life-cycle cost model. The user inputs for each case study are reported in Table 7.2.

Each of the case studies represents different scenarios for the proposed life-cycle cost model. However, case study A and case study D have obtained a six-star rating by Green Star. However, the proposed model supports only up to a five-star rating. Therefore, the case studies with six-star ratings are analyzed based on the upper boundary of the five-star rating level, which is 74 credits points. Two other case studies with five-star ratings are analyzed based on the mid-range boundary options available from the proposed model (refer Section 6.4).

7.3.6 Discussion on credit selection for case studies

All the case study buildings achieved almost all the credit points listed in "management" key criteria. This result is very similar to the output of the proposed model. In all the case study buildings "M3.0 Implementation of a Climate Adaptation Plan" is not addressed. However, this credit is selected by the proposed model in all the cases. The main reason for this is that this credit has two credit points attached to it. Therefore, the model considers it as an optimum selection because for a lower life-cycle cost, which is basically an initial cost, this credit provides two credit points in this key criterion. However, over 90% of management credits were achieved by all the case study buildings, including commissioning, building tuning, building a user guide, and having an environmental management plan. The results are similar. Most credits of the "management" key criterion are selected.

TABLE 7.2 User inputs for each case study.

User information/constraint	Case study			
	Case study A	**Case study B**	**Case study C**	**Case study D**
The desired level of green building certification using Green Star Design and As Built v1.1				
5 star rating	✓	–	–	✓
6 star rating	–	✓	✓	–
Central Business District (CBD) the building is located in				
Adelaide	✓	–	–	–
Brisbane	–	–	–	–
Melbourne	–	✓	–	–
Hobart	–	–	–	–
Perth	–	–	–	✓
Sydney	–	–	✓	–
Can the building users access the office location using public transport?	✓	✓	✓	✓
Are you willing to use at least 95% of all engineered wood products to meet the stipulated formaldehyde units? Or, are there no new engineered wood products used in the building?	✗	✗	✗	✓
Are you willing to source 95% of building steel from a responsible steel maker?	✓	✗	✓	✓
Are you willing to source 95% of timber used in the building with a forest certification scheme?	✓	✓	✓	✓

Continued

TABLE 7.2 User inputs for each case study—cont'd

User information/constraint	Case study			
	Case study A	Case study B	Case study C	Case study D
Are you willing to avoid using polyvinyl chloride (PVC) (at least 90%, by cost) and have an environmental product declaration (EPD)?	✗	✓	✓	✗
Are you willing to use an on-site renewable energy generation source?	✗	✓	✓	✗
Were there any critically endangered, endangered or vulnerable species, or ecological communities present on the site at the time of purchase?	✗	✗	✗	✗
Was the ecological value improved by the project?	✗	✓	✗	✗
Is the site classified as a site of "High National Importance?"	✗	✗	✗	✗
Is the site a previously developed land (at least 75%)?	✓	✓	✓	✓
Is the land previously contaminated and will the site be remediated with a best practice remediation strategy?	✗	✗	✓	✗
Are you willing to reduce the use of Portland Cement content in all the concrete used in the building by replacing it with supplementary cementitious material (SCM)?	✗	✗	✓	✗
Are you willing to use 95% of all internally applied paints, adhesives, sealants and carpets to meet the stipulated "total volatile organic compound (VOC) levels, or not to use paints, adhesives, sealants and carpets?"	✓	✓	✓	✓
Does the building has a clear line of sight to a high quality internal or external view?	✗	✓	✓	✓
Is the building located conveniently vis-à-vis amenities?	✓	✓	✓	✓
What is the type of ventilation system used the building?				
Mechanically ventilated air cooled system	–	–	–	–
Mechanically ventilated water cooled system	✓	✓	✓	✓

✓, denotes "yes" and ✗, denotes "no" in the proposed life-cycle cost model.

The IEQ key criterion is also widely achieved by all the case study buildings. Further, all these case studies achieved credit points such as "IEQ 9.2 Provision of Outdoor Air," "IEQ11.0 Minimum Lighting comfort," "IEQ11.1 General Illuminance and glare reduction," "IEQ 13.1 Paints, Adhesives, Sealants, and Carpets," "IEQ14.1 Thermal Comfort," and "IEQ14.2 Advanced Thermal Comfort." These credits perfectly coincide with the selection of credits by the proposed model. Case studies A and D achieve all the credits in the "Acoustic Comfort" sub-criteria. However, the buildings of case studies B and D do not consider achieving the credits related to acoustic comfort, although the proposed model suggested credits related to "Acoustic comfort" sub-criteria. The main reason is the lower life-cycle cost because it required only minimum regular maintenance (refer Section 5.5). The "IEQ12.2 Views" credit is achieved by all the case study buildings except by that of case study A. This credit only has an initial cost and it can be obtained to a certain extent through a better design. Therefore, the model proposed this credit for case study A, with a highlight to illustrate that there is a constraint involved in this credit. The building of case study D achieved the "IEQ13.2 Engineered Wood Products" credit by using the stipulated formaldehyde units. However, all the other case study buildings did not achieve this credit points. Once again, in terms of life-cycle cost, this credit has a lower life-cycle cost due to minimum regular maintenance. Therefore, the proposed model selected this credit for all the cases with highlights, illustrating the user constraints.

The "Energy" key criterion offers two main sub-criteria in Green Star rating, namely: "GHG emissions" and "Peak Electricity Demand Reduction." However, all the four case studies obtained credit points using the "15D GHG Emissions Reduction-NABERS Energy Commitment Agreement" pathway. This credit point is not included in this study. Therefore, this represents one of the main limitations of this study. Further, it is necessary to note that the proposed model always selected credits for on-site energy generation system. The model proposed this as the optimal suggestion in case studies B and C, and it is used in the corresponding buildings as well. According to Tam, Le, Zeng, Wang, and Illankoon (2017), using PV panels as renewable energy is a life-cycle cost saving which is similar to the life-cycle cost data in the proposed model. Therefore, the "E16A Prescriptive Pathway: On-site Energy Generation" credit is always given as a selection if there are no any user constraints.

All the case study buildings are located in the city itself. Therefore, all the buildings have access to public transport and are also closer to public amenities. Further, all the case study buildings have provided car parking spaces and cyclist facilities. Therefore, the credit points achieved by the case study buildings and the selections provided by the model perfectly overlap with each other.

In the "Water" criteria, the proposed model selected all the credits except for "W18B.3 Heat Rejection." This credit cannot be achieved because all the buildings used water for the heat rejection. Similar to the selection of the proposed model, all the buildings used water efficient sanitary fixtures. However, only the building of case study B used rainwater reuse initiative. This is selected by the model because of the water savings. Irrespective of the water savings in the life-cycle, the "W18B.2 Rainwater Reuse" credit was not considered by other case study buildings except by the one of case study B.

Using sustainably sourced timber and a responsible steel maker are common in the construction of green buildings. All the case studies achieved "Mat20.2 Timber Products" credits overlapping with those from the proposed model. Similarly, "Mat20.1 Sustainable Reinforcing Steel" is considered in all the case study buildings except for that of case study B. The model also has the same selection and for the building of case study B using steel for reinforcing steel has user constraints. Further, "Mat20.3 Permanent Formwork Pipes, Flooring, blinds and Carpets" is another credit selected by the proposed model unless there is a user constraint, and this selection always coincides with the credits achieved by the case study buildings.

All the buildings in the case studies have re-used previously utilized land. Further, except for case study B, all the other case studies had restrictions in developing the ecological value of the land. Therefore, only case study B achieved the "L23.1 Ecological Value" credit point. According to the proposed model, in an event of obtaining credit points, the "L23.1 Ecological Value" credit is selected and highlighted for the user constraint. The "Em29 Refrigerant Impact" and "Em26.1 reduced Peak Discharge" credit points are achieved by all the case study buildings and also selected by the proposed model.

For the "Energy" criterion, the model proposed on-site energy generation using photovoltaic panels. Although this initiative incurs a higher initial cost, according to the life-cycle cost, this gains a life-cycle saving for the project. If the initial decision making is based on the initial cost, this initiative will be eliminated, unless the building life-cycle is considered. Similarly, the rainwater reuse credit provides savings within the life-cycle, and using photovoltaics and reducing the heat island effect can be achieved together, as proposed by the model in certain instances.

In general, the credits selected by the proposed model and those achieved by the four case study buildings are similar to a greater extent. If a particular credit point results in savings over the life-cycle, the proposed model captures that in the calculation. Therefore, in such instances, there are differences between the credit selection of the proposed model and the case study buildings. As an example, credits such as "E16A Prescriptive

Pathway; On-site Energy generation" have massive savings when considering the life-cycle of the buildings. However, the initial cost of the particular credit point is higher. Therefore, the case study buildings, such as those of case studies A and D do not use renewable energy on-site. Considering the life-cycle savings, the model selects this credit unless there are not any user constraints. Further, if the building needs to further upgrade the level of certification, this credit will be selected while highlighting the user constraint by the proposed model so that the user can re-consider it further. Apart from that, the selections of credits with higher initial cost proportions and negligible proportions of operational, maintenance, and demolition costs (refer Chapter 5) are similar to those achieved by the four case study buildings. Therefore, the credits selected by the proposed model and those of the case study buildings are similar; hence, the model results are validated. Further, the proposed model provides optimum solutions considering the life-cycle savings, which are completely ignored by the case study buildings.

7.4 Summary

This chapter focused on validating the life-cycle costs and credit selections used by the proposed model. The costs used by the proposed model are within the range of costs obtained by different sources. The initial cost per square meter of the building in Sydney with air-conditioning and a five star rating is approximately 3250 AUD/m^2. This cost is checked against many databases and actual green star certified commercial office buildings. In all these cases, there were slight changes in the calculated costs, yet these were within an acceptable range. Further, all the costs are cross-checked by industry experts on a regular basis. Therefore, considering all the available data, this chapter concludes that the cost data used in the proposed model are accurate and consistent.

The other section of this chapter focused on the validation of the credit selection. For this, four case study buildings are considered. Each building has different user constraints and information. These different user inputs are given to the proposed model and the result is compared with the actual credits obtained by the buildings. According to the comparison, the credits obtained by the buildings and those selected by the proposed model perfectly overlap in most instances. However, the proposed model provides certain solutions considering the life-cycle cost savings, which are ignored by the certified buildings. Therefore, it is evident that this proposed life-cycle cost model provides solutions to the user considering the optimum selections that result in the lowest life-cycle cost and the highest credit point values.

References

Australian Institute of Quantity Surveyors. (2017). *Building cost index*. Sydney, Australia: Australian Institute of Quantity Surveyors.

Cordell. (2016). *Cordell cost guide for commercial building cost guide*. New South Wales (2016) ed *Vol. 47*. West Chatswood, NSW: Cordell.

Fellows, R. F. (2015). *Research methods for construction* (4th ed.). New York: Wiley.

Green Building Council of Australia. (2008). *The dollars and sense of green buildings*. Retrieved from Australia: https://www.gbca.org.au/resources/gbca-publications/dollars-sense-2008/.

Green Building Council of Australia. (2016). *Green Star financial transparency*. Retrieved from Australia: https://gbca-web.s3.amazonaws.com/media/documents/gbca-research-paper---financial-transparency-2016.pdf.

Rawlinsons. (2016). *Australian construction handbook. Vol. 34*. Perth, Western Australia: Rawlinson Publishing.

Tam, V. W. Y., Le, K. N., Zeng, S. X., Wang, X., & Illankoon, I. M. C. S. (2017). Regenerative practice of using photovoltaic solar systems for residential dwellings: An empirical study in Australia. *Renewable and Sustainable Energy Reviews*, *75*, 1–10. https://doi.org/10.1016/j.rser.2016.10.040.

Yin, R. K. (1994). Case study research: Design and methods. *Applied social research methods series: Vol. 5*. London: Sage Publications. Biograsphy.

CHAPTER

8

Conclusion

8.1 Introduction

This chapter discusses the major conclusions from this book, and details the recommendations derived from it. The study has certain limitations, which are also illustrated in this chapter. Finally, future research directions derived from the study, are provided in detail.

8.2 Conclusions of the study

"Green building" is a widely discussed concept in the construction industry, due to its environmental and social benefits. Although it is widely appreciated for its many advantages, higher initial costs have always been a hindrance to green building development. In the construction industry, usually, there are massive costs upfront and green buildings are considered to be higher in terms of initial costs, when compared to conventional buildings. However, the life-cycle savings of these buildings are rarely discussed, and a monetary evaluation of these savings have hardly been carried out. Therefore, there is a clear lack of research on life-cycle calculations in the implementation of green building construction. Considering this fact, this study proposed a life-cycle cost model, in order to select optimum credits, considering the lowest life-cycle cost for commercial office buildings in Australia, using Green Star Design and As-Built version 1.1.

Initially, all the credits of Green Star Design and As-Built version 1.1 were thoroughly analyzed, after which the life-cycle cost was calculated for each of the credits, and finally, the life-cycle cost model was proposed. This model considers all user constraints and uses various sources of information from the user, in order to adjust the life-cycle costs accordingly. Furthermore, it considers the inter-dependencies among credits and conditional requirements among other factors, in order to achieve an optimized solution. Based on the results, the following conclusions have been drawn.

https://doi.org/10.1016/B978-0-12-820062-9.00008-6

Chapter two of this book studies the literature on sustainable development, green buildings, and green building rating tools. According to the literature, there are numerous definitions that have been put forward to explain sustainable development. However, the definition given in the Brundtland Report is considered the most widely used definition of sustainable development. While illustrating the concept further, it was indicated that sustainable development has a threefold focus, namely, on environmental, social, and economic sustainability. Therefore, a holistic approach should be put forward to achieve sustainable development, fulfilling all the three aspects of sustainability.

Similarly, according to the literature (refer Sections 2.4 and 2.5), green buildings should satisfy the triple bottom-line constructs, namely, environmental, social and economic sustainability. Further, to evaluate these green buildings, green building rating tools have been developed, which act as yardsticks to measure green building initiatives. These green building rating tools have various key criteria (refer Table 2.1) and credits which are parameters to evaluate green buildings. Therefore, these key criteria represent the triple bottom-line. However, this requirement is not supported by most green building rating tools that exist today (refer Section 2.5). Further, the economic sustainability is mostly ignored in existing green building rating tools.

To ensure the desired outcomes of constructing green buildings, it is necessary to focus on all the key criteria. In certain rating tools, there is a significant emphasis on a single criterion. Therefore, there is a possibility of obtaining green certification by fulfilling only that one particular criterion, even though all the other key credit criteria are overlooked or even completely ignored (refer Section 2.5). Similarly, it is possible to obtain Green Star certification while completely disregarding one or more key criteria (refer Section 6.4). Therefore, the proposed life-cycle cost model was developed to focus on all the key criteria holistically.

According to Fig. 2.1, a clear majority of office buildings are gaining Green Star certification. However, there are other types such as public, healthcare, and industrial buildings, of which few have achieved Green Star ratings in Australia. There may be certain barriers preventing these kinds of buildings from obtaining green certification, and this should be analyzed further.

The initial cost premium varies based on the country in which research is carried out, the basis of comparison, the type of certification of the green building, and the type of building itself. However, with time, the initial cost premium tends to reduce. The main drawback in considering the initial cost is the inability to consider the life-cycle savings derived from green buildings.

There are many research studies that suggest life-cycle cost for green buildings (refer Section 3.3). The main reason for this suggestion is that by considering the life-cycle cost, it is possible to capture all the costs and savings over the life-span of the building. Further, according to the literature, green buildings should address the triple bottom-line sustainability. However, there is a clear lack of life-cycle cost models to identify optimum solutions, considering all three aspects of sustainability in green buildings.

This book carried out an in-depth analysis on the Green Star Design and As-Built version 1.1 credits. Based on the analysis there are inter-dependent credits in the Green Star rating (refer Section 4.4). Fig. 4.3 clearly illustrates the inter-dependencies among credits. These inter-dependent credits can influence each other. Therefore, these credits should be considered together, collectively, while obtaining green building certifications to achieve an optimum output. These inter-dependencies are embedded in the model based on a set of selection rules (refer Section 6.3.3). As an example, if a building chooses to be ventilated naturally, certain credits, such as "W18B.3 Prescriptive Pathway: Heat Rejection" and "Em28 Legionella Impacts from Cooling Systems" can be bundled together easily with the lowest cost, and awarded directly according to the Green Star Design and As-Built version 1.1. These solutions are coupled together in this model. Therefore, this model identifies all of these credits and provides optimal solutions, so that buildings can achieve green certification with simple and lower life-cycle cost initiatives.

Credits representing the IEQ criteria and material key criteria have the most number of inter-dependencies among credits. All the credits in material key criteria have inter-dependencies with other credits (refer Fig. 4.3). Therefore, during the material selection in the designing and construction stages, all these inter-dependent requirements must be specially considered. As an example, in certain cases, such as selecting steel, even the steel producer must be considered. Similarly, this is applicable for the IEQ credits, as well. As an example, while selecting paints for a green building, the designers must consider the VOC levels to fulfill certain credits in the IEQ criteria. Therefore, material selection in green building implementation is not a standalone process. It should be integrated with requirements given in other key criteria, especially pricing, which is a close consideration for the IEQ credits.

Chapter 5 of this book provides optimum solutions with lowest life-cycle cost and the highest number of credit points. Further, this chapter provides detailed information on life-cycle cost calculations and an analysis of each key criterion. Based on the analysis, discount rate can be identified as one of the significant parameters in life-cycle costing. The results of the proposed model are significantly influenced by the discount rate of the user. Credits with higher life-cycle cost contribute towards the costs incurred in the operational stage of building, and this is significantly impacted by the discount rate (refer Chapter 5). Usually, when the discount rate increases, the life-cycle cost decreases. However, when there are massive cost savings, life-cycle savings are higher when the discount rate is less. Based on sensitivity analysis, majority credits in key criteria, such as IEQ and energy, have a significant impact on the discount rate. Further, a couple of credits in the key criteria, such as water and emissions, have a slight influence on the discount rate.

Chapter 5 provides optimum solutions, considering the lowest life-cycle cost and highest credit point achievements. However, the optimum solutions provided in this chapter do not consider the inter-dependence of credits and user constraints. Therefore, the solutions are generic. Table 8.1 below provides a summary of the optimum solutions selected in Chapter 5.

While analyzing the solutions with lowest life-cycle cost in the ventilation system, all the credits selected naturally ventilated system as one of the optimum solutions. The main reason for this is the lower maintenance costs and life-cycle savings associated with natural ventilation. Other mechanically ventilated systems also gain energy savings throughout the life-cycle, but there is a significant maintenance cost involved throughout the life-cycle. Therefore, more passive solutions, such as using natural ventilation and natural daylight, should be considered in green building implementation.

In Chapter 5, each credit is analyzed considering the life-cycle cost obtained from different CBDs in Australia. There are slight changes in the life-cycle costs for a majority of the credits and there are considerable changes in the life-cycle cost for a few credits. The main reason for this is the change in initial cost, especially the labor rates, and the energy costs in most instances. However, the changes in life-cycle cost do not alter the optimum solutions considering the life-cycle cost. Land values are not included in this study, although they influence a couple of credits representing key criteria, such as transport, land use, and ecology. These land values are expected to vary significantly, across Australia. Therefore, if land prices are considered, there is a possibility to visualize certain changes in lowest life-cycle cost solutions.

There are many systems installed in green buildings providing various services. HVAC, electrical and water services are some of the few, yet major, services within a building. There are a lot of credit points that are governed directly by these systems. Further, these credits represent various key criteria (refer Fig. 4.3). Usually, these services are provided by specialized contractors. Therefore, the requirements for each credit must be communicated through necessary specifications to the specialists.

Location and the type of land are among the important factors in green building implementation. The location of the land directly affects the transport key criteria of the green building. Further, 6 out of 20 questions developed for capturing user information and constraints focus on the location, type, and status of the land (refer Table 6.1). Selection of land directly influences many user constraints. Therefore, the developer must

TABLE 8.1 Summary of solutions for each key criterion.

Management	**IEQ**	**Energy**	**Transport**
• M1.0 Accredited Professional • M2.1 Services and Maintainability Review • M4.1 Building Operations and Maintenance Information • M4.2 Building User Information • M5.1 Environmental Building Performance • M5.2 End of Life Waste Performance	• IEQ12.1 Daylight • IEQ9.2 Provision of Outdoor Air—Naturally Ventilated • IEQ10.1 Internal Noise Levels • IEQ10.2 Reverberation • IEQ10.3 Acoustic Separation	• E16A Prescriptive Pathway: On-site Energy Generation • E15A GHG Emissions Reduction—Naturally Ventilated • E15A GHG Emissions Reduction—Domestic • Hot Water System • E15A GHG Emissions Reduction—Lighting	• T17B.1 Access by Public Transport • T17B.5 Walkable Neighborhood • T17B.4 Active Transport Facilities • T17B.2 Reduced Car Parking Provisions
Water	**Material**	**Land use and ecology**	**Emissions**
• W18B.3 Heat Rejection—Naturally Ventilated • W18B.4 Landscape Irrigation • W18B.2 Rainwater Reuse • W18B.1 Sanitary fixture efficiency	• Mat21 Product Transparency and Sustainability • Mat22A Reduction of Construction and Demolition Waste • Mat20.3 Permanent, Formwork, Pipes, Flooring, Blinds, and Cables • Mat20.2 Timber Products • Mat20.1 Structural reinforcing Steel	• "L23.1 Ecological Value" • "L24.1 Reuse of Land" • "L25 Heat Island Effect Reduction"	• Em28 Legionella Impacts from Cooling Systems—Naturally Ventilated/Air Cooled System • Em29 Refrigerant Impact—Naturally Ventilated • Em27.1 Light Pollution to Night Sky • Em26.1 Reduced Peak Discharge • Em26.2 Reduced Pollution Targets

thoroughly evaluate the plot of land prior to purchase, because once the plot of land is purchased, these constraints cannot be eliminated.

Certain credits with lower life-cycle costs are mostly eliminated from the initial decision-making stages owing to higher initial costs. According to the literature (refer Chapter 2) and many research studies (Davis & Langdon, 2007; Gabay, Meir, Schwartz, & Werzberger, 2014; Green Building Council of Australia, 2016; Kim, Greene, & Kim, 2014; Tatari & Kucukvar, 2011), there is a considerable initial cost premium in green building implementation. These studies have only focused on the initial cost of the green building which is the same situation in the actual green building decision-making stage. However, when compared with the proposed life-cycle cost model, there are credits with lower life-cycle costs that have been eliminated while achieving credits with comparatively higher life-cycle costs (refer Section 7.3.6), because only the initial cost is analyzed during the initial decision-making stages, completely ignoring life-cycle savings. For example, using PV panels can be a life-cycle cost-saving measure, but, it is ignored during the initial stages as a result of the higher initial cost.

For the cost validation of the research, Chapter 7, quotes approximate initial costs of green buildings after considering cost databases, industry reports, and an actual green certified building. Based on these facts, the initial cost of a green building in Australia approximately varies from a minimum of 2970 AUD/m^2 to a maximum of 3627 AUD/m^2.

The proposed model always selects a majority of the credits in management key criteria (refer Section 7.3.6). Further, according to the inter-dependencies that have been identified, all the credits in management key criterion are standalone credits with minimum or no inter-dependency among credits in other key criteria (refer Fig. 4.3). According to the literature, management is a function that occurs throughout the different stages of the green building life-cycle. Therefore, while developing a green building, users can initially look into credits in the management key criterion easily, because it has a lower life-cycle cost and these credits mostly act as support functions to green building implementation, as well.

IEQ criterion is widely achieved by green certified buildings. Credit points related to provisioning of outdoor air, lighting comfort, thermal comfort, and selecting material with low VOCs are significantly used in certified buildings. However, it is necessary to note that, even the lowest life-cycle cost solutions focused on natural ventilation and acoustic comfort, are rarely used on certified buildings. The main reason for this is the lack of information on life-cycle perspectives in the initial decision-making stages. These can be eliminated by using the proposed life-cycle cost model in the initial decision-making stages.

Unless there are user constraints, the proposed model is always set to use PV panels as a renewable energy generating source in green buildings. The main reason for this is significant life-cycle cost savings. However, certain green buildings use this approach whereas certain other buildings do not consider this option.

Apart from that, all the case study buildings used NABERS rating scheme to obtain energy credits, which are not discussed in this study. NABERS is a rating tool that evaluates the performance of given criteria, such as energy and water, separately, and provides certification considering the operational phases of the building (Illankoon, Tam, Le, & Tran, 2017). Therefore, to capture these possibilities, the life-cycle cost model can be developed further by integrating these tools as well.

This study focuses on commercial office buildings. Therefore, most of the buildings that fall into this category are located in the main CBDs with close proximity to public transport. As a result, the credits selected by the proposed model and the credits achieved by the case study of buildings coincide with each other. If other office buildings are considered, this can be similar.

Using water efficient sanitary fixtures is one of the most common practices nowadays in buildings. However, all the case study buildings used mechanically ventilated systems and therefore, "W18B.3 Heat Rejection" credit was eliminated. However, the proposed model always suggests natural ventilation to buildings, which makes this credit directly achievable. This is an optimum solution, always selected by the proposed model, unless there are user constraints for ventilation. This selection further suggests the usage of passive solutions, such as natural ventilation in green buildings, because it can obtain multiple benefits. Rainwater reuse is also a similar solution selected by the proposed model.

In almost all the case studies, material was sourced from a responsible supplier. Further, material was selected by the proposed model unless there was a constraint. Similarly, according to the validation, the credits selected by the proposed model coincided with the selections of credits by the green buildings for material credits. Therefore, it is possible to conclude that material selection in green building implementation focuses on the required standards practiced in the industry.

Land use and ecology key criterion has many user constraints. However, since this research considers commercial office buildings, usually, the location of the land is in close proximity to a CBD. Therefore, most of the available lands are previously developed lands, because it is highly unlikely to obtain an undeveloped plot of land within a CBD. As a result, credits related to re-using a previously developed land were achieved in all the case studies. The proposed model also includes this. As mentioned earlier, these criteria depend on many user constraints and the life-cycle cost should also consider the land value which is not considered in this study.

The credits selected by the proposed model and the credits obtained by the certified green buildings coincide to a great extent in in emissions key criterion. Credit such as "Em26.1 Reduced Peak Discharge" is popular in green building implementation and has been proposed by the life-cycle cost model as well.

Green buildings with Green Star ratings achieved credits with lower life-cycle costs, as proposed by the life-cycle cost model. However, many options exist for further improvement of green building ratings with minimum life-cycle costs. In comparing the results of the proposed model and the four case study buildings, the life-cycle cost model has suggested certain options that are not considered by the user, having proposed them as alternative solutions. Therefore, these suggestions can be used to further improve the green building status of a particular building. In other words, using this proposed model, users can identify the constraints that need to be eradicated to achieve a higher green certification. This concludes the possibility of using the proposed model to identify the next best solutions to increase the levels of certification of the green building using Green Star Design and As-Built version 1.1.

8.3 Recommendations to the green building industry

Based on the conclusions, it is evident that certain optimum credits with lower life-cycle costs are eliminated due to higher initial costs during the initial decision-making stages. Therefore, this study clearly signifies the importance of looking into life-cycle cost perspective, while deciding on the green building implementation within the initial decision-making stages. Further, this study makes a case for the use of the proposed life-cycle cost model to identify optimum credit selections while developing Green Star certified commercial office buildings.

Further, the authors propose this life-cycle cost model to be used to identify the next best solutions to be implemented to achieve a higher Green Star certification level. For this purpose, the users can first set the user constraints in the user inputs and set the required Green Star certification to a higher level. Then the proposed model will identify constraints that need to be eradicated to achieve a higher Green Star certification by addressing red flagged suggestions.

According to Green Star Design and As-Built version 1.1, it is possible to achieve Green Star rating while completely ignoring one or more key criteria. In such a scenario, the building that is certified as a Green Star certified building does not uphold the true green building requirements. Therefore, there is a clear need to set up mandatory requirements in Green Star Design and As-Built version 1.1 to achieve a minimum number of credit points in each of the key criteria prior to Green Star certification.

Discount rate used to calculate the life-cycle cost has significant importance. It directly influences the selections based on life-cycle cost.

Therefore, a user should be more careful and accurate in deciding the discount rate in life-cycle cost calculations. The same applies to the proposed life-cycle cost model as well.

Green building implementation is not initiated from the design stage of the building. According to the conclusions, the selection of the plot of land also needs to be considered carefully while developing a green building. Therefore, the benefits of green buildings and perceived life-cycle cost savings need to be communicated to the public, so that even prior to the design stage, the developers might consider developing a green building at the inception stages.

According to the conclusion of this book Green Star Design and As-Built version 1.1 do not include credits to evaluate the economic sustainability of green buildings. It is clear that this parameter of sustainability is overlooked by this green building rating tool. Therefore, the author recommends including certain credits to include the attributes of economic sustainability, while developing the Green Star Design and As-Built version 1.1 rating tools further.

Management credits act as support functions for green building construction. Further, to obtain the credits representing the material criteria, the suppliers also should be selected properly. Specialist contractors providing the HVAC and electrical systems must also be aware of the perceived green building certification. Therefore, in summary, the green building implementation should be a process that involves various professionals attached to various stages of the building's life-cycle. As a result, this study recommends looking into green building implementation as a whole process, integrating all these requirements.

The proposed life-cycle cost model always selected passive solutions, such as using natural ventilation, daylight in buildings, and rainwater tanks as optimum solutions. However, provisions such as natural ventilation are rarely used in green buildings. Therefore, it is recommended to consider these passive solutions in green building implementation.

Further, while considering Green Star certification, based on the type of building, certain building types are still lagging behind. Therefore, these types of buildings should be identified, and developers should be encouraged to achieve green building certification.

The proposed life-cycle cost model is recommended for various stakeholders such as policy makers, clients, designers and consultants. Policy makers such as the Green Building Council of Australia can use the conclusions of this study to further develop the rating tools. Further, clients, designer and consultants can used the proposed life-cycle cost model to identify the life-cycle costs and the various options available to achieve Green Star rating. The proposed life-cycle cost model is highly recommended to designers and consultants for regular use in developing green buildings from the initial decision making stages. These professionals can effectively identify the optimum solutions in terms of life-cycle cost and provide effective advice to clients.

8.4 Limitations of the study

This study proposes a life-cycle cost model for optimum selection of Green Star credits for commercial office buildings using Green Star Design and As-Built version 1.1. However, there are limitations to this research, which are as follows:

a. This study eliminated certain credits due to given reasons (refer Section 4.3.8). Therefore, only 74 credit points out of 100 credit points were considered for the study.
b. There are many green building rating tools used worldwide. However, the proposed model focuses only on Green Star Design and As-Built version 1.1 rating tools published by the Green Building Council of Australia.
c. This proposed model can be used to select optimum solutions for four star and five-star green buildings only.
d. All the life-cycle cost calculations are developed considering a commercial office building. Therefore, using this proposed model for other types of buildings will not be completely accurate.
e. All the cost data and construction details for this proposed model are collected in the Australian context. Further, all the costs are given in Australian Dollars.
f. While calculating costs using first principles, certain sensible assumptions were made. They are clearly mentioned in the study and in the proposed model.
g. This model focuses only on concrete framed buildings, and steel structures are not considered in the study.
h. The life-span considered for the green building is 60 years.
i. The proposed model does not provide an option to calculate and compare life-cycle costs for innovative solutions. This model only focuses on the given credits.

8.5 Future research directions

The proposed life-cycle cost model can be replicated to cater to different types of buildings. This proposed model only focuses on commercial office buildings, and can be further extended to other types of buildings, such as residential buildings, hospitals, and so on.

In this study, the focus was only on Green Star Design and As-Built version 1.1. However, this proposed model can be extended to suit other green building rating tools used worldwide.

This study eliminates all the prescriptive pathways that were allowed to achieve credit points. Prescriptive pathways encourage using various practices to achieve a given standard. Therefore, there can be various methods available to achieve these credits. The authors suggest future research to evaluate and identify the optimum solutions for each of these prescriptive pathway credits.

The land value changes significantly across main CBDs and areas within Australia. Therefore, the cost premiums included in obtaining a land in a prime location are not included in the proposed model. Therefore, future research can be undertaken to analyze and include premium costs that the developers should pay to obtain plots of land and embed this into the proposed life-cycle cost model.

There are other green building rating tools such as NABERS and NatHERS that can be used to achieve credits in Green Star Design and As-Built version 1.1. These rating tools can also be embedded into this model. Therefore, future research can be directed focusing on integrating other rating tools used for certification, as well.

Green building implementation should be considered a process and all the other functions, such as supplier selections and specialist subcontractor works, should be integrated. Therefore, future research can be directed to develop a model to integrate all these functions to ensure smooth delivery of green buildings.

Although the proposed model identifies passive methods, such as natural ventilation to green buildings, these options are rarely practiced. Therefore, future research can be carried out to identify the reasons for this and identify ways to encourage developers to use these methods.

There are certain drawbacks in Green Star Design and As-Built version 1.1 as identified in this book. Therefore, future research can be undertaken to further develop and promote Green Star Design and As-Built green building rating tool.

There are other Green Star rating tools focusing on building and green neighborhood such as Green Star—Interiors, Green Star—Communities, and Green Star—Performance. Future research can be conducted to develop life-cycle cost models for these rating tools, so that the life-cycle cost benefits can be communicated to the community, in a much broader manner.

There are certain building types, such as health care buildings and industrial buildings, which do not seek green building certification often. There can be certain barriers which prevent these buildings from obtaining green certification. Therefore, future research can identify those barriers and propose solutions to enhance green certification.

References

Davis, & Langdon. (2007). *The cost and benefits of achieving green buildings*. Retrieved from http://www.usgbc.org/Docs/Archive/General/Docs2583.pdf.

Gabay, H., Meir, I. A., Schwartz, M., & Werzberger, E. (2014). Cost-benefit analysis of green buildings: An Israeli office buildings case study. *Energy and Buildings, 76*, 558–564. https://doi.org/10.1016/j.enbuild.2014.02.027.

Green Building Council of Australia. (2016). *Green star financial transparency*. Retrieved from Australia: https://gbca-web.s3.amazonaws.com/media/documents/gbca-research-paper---financial-transparency-2016.pdf.

Illankoon, I. M. C. S., Tam, V. W. Y., Le, K. N., & Tran, C. N. N. (2017). Review on green building rating tools used in Australia. In *Paper presented at the international conference on innovation production and construction (IPC), Perth, Australia*.

Kim, J.-L., Greene, M., & Kim, S. (2014). Cost comparative analysis of a new green building code for residential project development. *Journal of Construction Engineering and Management, 140*(5). https://doi.org/10.1061/(ASCE)CO.1943-7862.0000833. 05014002-05014001–05014002-05014010.

Tatari, O., & Kucukvar, M. (2011). Cost premium prediction of certified green buildings: A neural network approach. *Building and Environment, 46*(5), 1081–1086. https://doi.org/10.1016/j.buildenv.2010.11.009.

Index

Note: Page numbers followed by *f* indicate figures and *t* indicate tables.

M

N

P

R

S

T

U

V

W

9780128200629